NEBS MANAGEMENT DEVELOPMENT
SUPER SERIES
THIRD EDITION
Managing Activities

Achieving Quality

Published for
&NEBS Management by

Pergamon
Open
Learning

Pergamon Open Learning
An imprint of Butterworth-Heinemann
Linacre House, Jordan Hill, Oxford OX2 8DP
A division of Reed Educational and Professional Publishing Ltd

 A member of the Reed Elsevier plc group

OXFORD BOSTON JOHANNESBURG
MELBOURNE NEW DELHI SINGAPORE

First published 1986
Second edition 1991
Third edition 1997

© NEBS Management 1986, 1991, 1997

All rights reserved. No part of this publication may be reproduced in any material form (including photocopying or storing in any medium by electronic means and whether or not transiently or incidentally to some other use of this publication) without the written permission of the copyright holder except in accordance with the provisions of the Copyright, Designs and Patents Act 1988 or under the terms of a licence issued by the Copyright Licensing Agency Ltd, 90 Tottenham Court Road, London, England W1P 9HE. Applications for the copyright holder's written permission to reproduce any part of this publication should be addressed to the publishers

British Library Cataloguing in Publication Data
A catalogue record for this book is available from the British Library

ISBN 0 7506 3297 6

The views expressed in this work are those
of the authors and do not necessarily reflect
those of the National Examining Board for
Supervision and Management or of the publisher.

NEBS Management Project Manager: Diana Thomas
Author: Joe Johnson
Editor: Fiona Carey
Series Editor: Diana Thomas
Based on previous material by: Joe Johnson
Composition by Genesis Typesetting, Rochester, Kent
Printed and bound in Great Britain

Contents

Workbook introduction   v
1. NEBS Management Super Series 3 study links   v
2. S/NVQ links   vi
3. Workbook objectives   vi
4. Activity planner   vii

Session A   Quality in context   1
1. Introduction   1
2. The meaning of quality   1
3. Ensuring quality   4
4. Total quality management   7
5. Achieving quality at team level   9
6. **Summary**   13

Session B   Standards   15
1. Introduction   15
2. National and international standards   15
3. Product standards and quality systems standards   16
4. The quality systems standard BS EN ISO 9000   17
5. **Summary**   30

Session C   Quality control and statistics   33
1. Introduction   33
2. The average or mean   33
3. The range   35
4. The standard deviation   37
5. The distribution of data   41
6. **Summary**   51

Session D   Statistical process control   53
1. Introduction   53
2. Sampling   53
3. Probability   54
4. Acceptable quality level (AQL)   60
5. Control limits   63
6. Applying the techniques   75
7. **Summary**   80

Performance checks   81
1. Quick quiz   81
2. Workbook assessment   83
3. Work-based assignment   84

Reflect and review 87

1 Reflect and review 87
2 Action plan 89
3 Extensions 91
4 Answers to self-assessment questions 92
5 Answers to activities 95
6 Answers to the quick quiz 97
7 Certificate 98

Workbook introduction

1 NEBS Management Super Series 3 study links

Here are the workbook titles in each module which link with *Achieving Quality*, should you wish to extend your study to other Super Series workbooks. There is a brief description of each workbook in the User Guide.

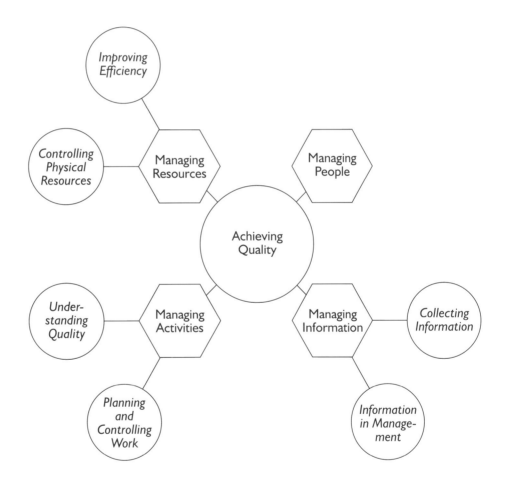

Workbook introduction

2 S/NVQ links

This workbook relates to the following elements:

- A1.1 Maintain work activities to meet requirements
- A1.3 Make recommendations for improvements to work activities
- C12.1 Plan the work of teams and individuals

It is designed to help you to demonstrate the following Personal Competences:

- analysing and conceptualizing;
- building teams;
- focusing on results;
- thinking and taking decisions;
- striving for excellence.

3 Workbook objectives

In a more competitive world organizations will only survive if they can guarantee quality in their goods or their services. Short-term profit at the expense of quality will lead to short-term lives. In that sense quality is, to my mind, the organizational equivalent of truth. Quality like truth will count, in the end. No one, and no organization, can live a lie for long.[1]

Does your organization try to cover up the truth? Does it pay lip-service to quality? Or do you believe it has quality so firmly in its sights that it has nothing to hide?

This workbook is about quality in organizations. You should find it interesting: it encompasses a broad spectrum of ideas and aspects of the subject.

There are four sessions. Session A is intended to be a general discussion of quality: what it is, what it's for, and how it can be achieved. Having set the scene, in Session B we home in on standards, and specifically ISO 9000, the quality systems standard. We'll look in some detail at what this standard contains.

Sessions C and D are rather more technical: they deal with statistics, and statistical process control (SPC). We will attempt to answer the question: 'Having set up your work process to achieve a defined level of quality, how can you keep it under control?'

---

[1] Charles Handy, *The Age of Unreason*, p. 115. Arrow Books Ltd.

3.1 Objectives

When you have completed this workbook you will be better able to:

- explain the meaning and purpose of quality;
- describe some sound approaches to quality management;
- summarize the contents and purpose of ISO 9000;
- carry out simple statistical and probability calculations related to quality control;
- recognize how the techniques of statistical process control can be usefully applied to work processes.

4 Activity planner

You may want to look at the following Activities now, so that you can start collecting material as soon as possible.

> Activity 4 where you are asked to recommend a way to improve your team's performance.
>
> Activity 8 which asks you to make a plan for at least one way of improving the extent to which your team's performance complies with the Quality Manual. If you aren't already familiar with the Quality Manual, you may want to set aside time to read it.
>
> Activity 9 where you are expected to set out an action plan or proposal for improving performance in controlling work processes.

Some or all of these Activities may provide the basis of evidence for your S/NVQ portfolio. All Portfolio Activities and the Work-based assignment are signposted with this icon.

The icon states the elements to which the Portfolio Activities and Work-based assignment relate.

In the Work-based assignment you are given the choice of (a) finding ways to apply statistical process control to your team's work or (b) selecting an area of quality management, and drawing up a brief report which explains how your team can contribute more effectively to the organization's quality system or procedures. Both these tasks are designed to help you meet elements A1.1 and A1.3; they will also contribute to elements D1.1 and D1.2 of the MCI Management Standards: 'Gather required information', and 'Inform and advise others.' You may want to prepare for the assignment in advance.

Session A Quality in context

1 Introduction

> The heart of quality is not a technique. It is a commitment by management to its people and product – stretching over a period of decades and lived with persistence and passion – that is unknown in most organizations today.[1]

EXTENSION 1
The Fundamentals of Quality Management, by Dennis F. Kehoe, covers all aspects of this workbook in some depth.

The book by Peters and Austin was first published in 1985. There's been a growing interest in quality management since that time, but if the criticism implied in the second sentence was valid then, the situation won't have changed very much – there hasn't been time. If you are lucky, you may have noticed a maturing commitment to quality among senior management in your organization. At any rate, you will perhaps be aware of the increased importance given generally to the management of quality.

The last two sessions of this workbook are about techniques: statistics, sampling and quality control. However, as Peters and Austin say, techniques are not at the heart of quality. It is important, therefore, to begin by discussing some more fundamental issues, before we get on to numbers and statistics.

The Super Series 3 workbook *Understanding Quality* would be a good place to start if you want to learn more about the subject of quality in general.

We start by defining what we mean by quality. Then we discuss the management aspects, and the steps involved in achieving high quality.

We'll round off the session by briefly investigating the subject of Total Quality Management, and achieving quality at team level.

2 The meaning of quality

When you want to buy something – a radio, say, or a haircut – there's usually a choice of suppliers, and brands or styles. So even if you have a very clear idea of what you're looking for, a decision has to be made.

2.1 The starting point for quality

It is the decisions made by customers like you that determine the quality of goods and services.

---

[1] Tom Peters and Nancy Austin, writing in *A Passion for Excellence*, p. 101. First published in the USA by Random House (1985). First published in Great Britain by William Collins: Fontana/Collins (1985).

Session A

Activity 1

Suppose you intend to purchase a fairly expensive item of household equipment – say a washing machine or vacuum cleaner.

What would influence the choice you make? Jot down **four** factors that would influence your choice.

_____

_____

_____

_____

There may be a number of factors influencing your decision, including perhaps:

- your experience when using similar products in the past;
- the price: how much you are prepared to pay, and how one model compares in price with another;
- the features available on each model;
- a recommendation of a particular product by a friend;
- what machines a shop has in stock;
- the colour and appearance of the product;
- whether the manufacturer has a reputation for the reliability of its products;
- delivery (can you have the product when you need it?);
- the persuasiveness of the shop salesperson;
- whether the goods meet agreed safety standards;
- the size of the product (will the machine fit in your kitchen?);
- and so on.

All these factors have some bearing on the quality of the goods. Your final choice of product will differ from that of many other people because you have different needs from them. Producers and suppliers have to take careful note of the choices made by potential customers. If they try to market a product that people don't want, they are bound to fail.

The starting point for quality is in the wants and needs of customers.

Session A

2.2 The definition of quality

What exactly do we mean by the word quality?

British Standard 4778 (the glossary of terms used in quality assurance) defines quality as follows:

The totality of features and characteristics of a product or service that bear on its ability to satisfy stated or implied needs.

This has been restated in a number of ways. One way to summarize it is to say that:

Quality is fitness for purpose.

Activity 2

A book is published first in hardback form, with superior binding and on expensive paper. Later the same book is published as a paperback, on cheap paper.

Which book, if any, is of the higher quality? Give a brief reason for your answer.

_____

_____

_____

You may have said that the hardback book was of higher quality. This would be a common understanding of the word 'quality'. A hardback book feels and looks better, and an advertisement describing the hardback book might talk about 'quality paper' or 'quality materials'.

> A producer (such as a bicycle manufacturer, or a book publisher) generates goods or services for sale. A supplier (such as a flower shop, or an insurance agent) offers goods or services for sale.

But we have said that quality is fitness for purpose. If a product becomes too expensive for the customers it is aimed at, because inappropriate materials are used, it won't sell. The hardback book will suit certain customers, but many others would sooner wait for the paperback. Both products are designed with the needs of their intended customers in mind. Neither is inherently of higher quality than the other.

So every producer and supplier has to make a decision as to the quality required by customers. If the quality of a product is not appropriate for their needs and wants – using materials that are too expensive, perhaps, or too cheap – people will not buy it (or will not continue to buy it).

Session A

3 Ensuring quality

If quality is determined by the needs, wants or expectations of customers, or potential customers, how can a producer or supplier go about meeting those needs?

The process can be expressed in terms of four steps: four questions that the organization must answer:

> From now on, we'll use the word 'product' to include any kind of goods or services. This definition is supported by ISO 9000, as we will discuss later.

1 What do customers need or want?

2 Are our products meeting those wants?

3 How can we specify new or modified products to meet those wants?

4 How can we make sure that our products match, and continue to match, our own specifications?

The first two steps are in the domain of marketing, and the questions can be answered by conducting activities such as:

- carrying out market research;
- talking to customers and potential customers;
- studying competitors' products;
- analysing product performance;
- investigating technological developments;
- keeping abreast of fashion trends.

Answering the third and fourth questions requires an understanding of design quality and process quality. These are defined as follows:

Design quality: the degree to which the specification of the product satisfies customers' wants and expectations. (Is it the right product?)

Process quality, or the quality of conformance: the degree to which the product conforms to specifications, when it is transferred to the customer. (Are we producing it right?)

In Sessions C and D, our focus will be on process quality. We will look at some statistical techniques that are used in controlling the quality of goods and services. Design quality is equally important, as we will discuss in Session B.

3.1 Being precise about quality

However, it is important to remember that you can control quality only after you have defined the standards you want to reach. It's not enough to talk in vague generalities, encouraging employees to 'use high quality materials', or

Session A

'work to the very best of your abilities'. Getting quality right means being very specific and detailed about how you want your product to perform: how you intend to meet your customers' needs.

Any reputable manufacturer will go to great lengths to specify dimensions, materials, construction methods, test methods, performance criteria and so on.

A service provider may find it more difficult to be so explicit about the standards of quality expected. It takes more thought if you want to give detailed instructions to sales staff about the level of friendliness required, or give precise guidelines to nurses regarding the amount of attention they should give to patients in particular circumstances. Although these are difficult questions, they must be addressed.

What we mean by quality in any complex organization is likely to be determined by a whole range of judgements. Many organizations are continuing to put a lot of effort into setting standards that are not easily measured in terms of output or defect rates.

3.2 Getting the quality right

Who is responsible for the quality of the product or service with which you are associated?

Activity 3

Which people in your current (or a previous) organization are responsible for ensuring the quality of its products?

The teams of people who do the work ☐

The people who check the work ☐

First line managers ☐

Top management ☐

All of the above ☐

The answer can be found on page 95.

The quality of a company's products and services is a reflection of the organization of that company and of the training and attitudes of its management and employees.

5

Session A

Getting the quality right means:

- **A clear commitment from higher management**

  If management are not interested in quality, then they should not be surprised if the company gets a reputation for poor quality.

- **The setting of well-defined standards**

  Everyone needs to know what he or she is aiming at. If standards are unclear, there is bound to be great variability.

- **Providing the resources that will enable those standards to be reached**

  The word 'resources' includes: workspace, materials, equipment, finance and training.

- **Allowing and encouraging employees to be responsible for the quality of their own work**

  Getting the quality right means not only setting standards, but giving people the opportunity to take full responsibility for meeting those standards.

  The old idea of making those in the production department responsible for production targets, and making the quality control department responsible for quality targets, has been shown not to work.

- **Setting a culture for quality**

  It is a fact of life that most people tend to set their standards by the behaviour of those around them. If management behave as if they care about quality, the rest of the staff will be much more inclined to follow suit.

- **Being consistent**

  Not all suppliers are clear what quality standards they expect their products to reach.

- Tanya Shipley went into a high street store on Wednesday, and was served by a courteous, helpful and well-informed assistant. Feeling happy about the service she got, Tanya returned to the same store on Saturday, to make another purchase. This time, once she managed to get the attention of a member of staff, she found him to be surly, lacking in knowledge about the product, and off-hand in answering her questions. When she complained, Tanya got an apology from the manager, who said: 'Oh, he is temporary Saturday staff – what can you expect?'

> **Consistency is a key aspect of quality.**

What Tanya expected was a consistently high level of service. As a customer, she was not very interested in the problems the store had in achieving this standard of quality.

- **Ensuring that quality standards are adhered to**

  This is what is usually meant by **quality assurance** and **quality control**.

Session A

3.3 Quality assurance and quality control

The definition of **quality assurance** is

All those planned and systematic actions necessary to provide adequate confidence that a product or service will satisfy given requirements for quality. (BS 4778)

Quality control is part of quality assurance. It consists of the operational techniques and activities that are used to fulfil requirements for quality, and which are the focus of the last two sessions of this workbook.

Let's repeat the main points of this section.

- Producers and suppliers need to know their customers' wants, and whether their products are meeting those wants.
- They must then specify the goods and services they intend to produce or supply: that's **design quality**.
- The next stage is to ensure that the defined quality standards are adhered to: that's **process quality**, and it's achieved through systems of **quality assurance** and **quality control**.
- Quality is the responsibility of everyone in the organization.
- Achieving quality depends on management:
    - displaying commitment;
    - setting standards;
    - providing resources;
    - allowing employees to take responsibility for standards;
    - setting a culture for quality.

4 Total quality management

In the long term, organizations compete on the basis of their product or service quality; TQM is therefore seen as an approach to gaining or sustaining a competitive lead.[2]

Achieving quality, including getting accreditation to ISO 9000 or other standards, is no easy journey to make.

Experience shows that a piecemeal approach is ineffective. Organizations that tackle quality problems in isolation, or work on the assumption that quality is some magical ingredient to be sprinkled on to taste, generally come to grief.

[2] Dennis F. Kehoe in *The Fundamentals of Quality Management*, p. 89. Chapman and Hall. First edition (1996).

Session A

Quality is not separate from the product: it is an integral part of the whole and all its parts. It follows that you cannot 'add on quality'; neither can you 'inspect in quality'. Quality must be:

- built in through every stage;
- intrinsic to every process;
- an essential element within every component.

The implication, therefore, is that quality must be total, and must involve every person and every activity. The term generally used for this 'holistic' approach is **Total Quality Management (TQM)**.

We can define TQM as follows.

Total Quality Management involves every member of the organization in a process of continuous improvement with the aim of satisfying the customers' wants and expectations.

The aims of quality, as we have already agreed, are to satisfy the customer. What about continuous improvement?

4.1 Continuous improvement

Continuous improvement is often referred to by its Japanese name, **kaizen**. It means carrying out many (perhaps small) detailed improvements to products, procedures and practices, over a long period.
Teams and individuals are expected to search for better ways of doing things, and for higher standards.

Some examples of the way things might be improved are:

- putting by-products to good use, instead of throwing them away;
- analysing the order in which the steps of a process should be carried out, so as to find the most efficient and effective sequence;
- finding better ways to communicate detailed instructions to individuals;
- reorganizing a team, so that each member gets the chance to make better use of his or her skills and knowledge;
- encouraging people to talk more about mutual problems;
- setting aside prejudices;
- eliminating unnecessary bureaucracy;
- identifying ways of making the job more interesting and less of a drudge.

Session A

 Activity 4

15 mins

This Activity is the first of a series of three which together may provide the basis of appropriate evidence for your S/NVQ portfolio. The others are Activities 8 and 9. If you are intending to take this course of action, it might be better to write your answers on separate sheets of paper.

Your aim in this Activity is to recommend a way to improve your team's performance.

Remember that, as a manager, you are most effective when you are able to give your team members the opportunity to make their own improvements. So one recommendation or action by you should, ideally, trigger off a whole series of improvements.

You may want to spend some time thinking about your team's work: their workspace; the pressures on individuals; their efficiency; their performance in comparison with other teams; their morale; their powers to take the initiative.

When you have identified a specific means of effecting improvement, write it down in detail on a separate sheet of paper.

5 Achieving quality at team level

If you are a first line manager working for a producer or supplier, you have a responsibility for the work done by your team, and for the quality of the goods and services you provide.

Session A

Activity 5

How do you ensure that the required levels of quality are met? Tick whichever of the following you consider to be important in achieving quality in your job, and then say how each action is implemented in practice.

What do you do?	How do you do it?
■ I let the team know the standards of quality expected by me and the organization. ☐	
■ I demonstrate that I am just as concerned about quality as meeting other team goals, or controlling costs. ☐	
■ I take complete responsibility for the quality of my team's work ☐	
■ I take steps to ensure that training and other resources are provided, so that my team members are able to perform effectively. ☐	
■ I encourage the team in their efforts to raise standards. ☐	
■ I check the team's work while it is in progress. ☐	
■ I check the quality of the completed work (perhaps by sampling a percentage of the output). ☐	
■ I check just about everything that every team member does. ☐	

Session A

Let's discuss each of these points.

- **Letting the team know the standards of quality you and the organization expect**

  This is certainly important, as we've already mentioned. Even when quality standards have already been set by others – designers, managers or other departments – the first-line manager retains responsibility for:

  - making sure the standards are understood;
  - imposing higher standards than the minimum, wherever possible;
  - leading the way by personal example.

  Part of your crucial role is to help get the quality message across.

- **Demonstrating a concern for quality**

  Although this point seems similar to the last, there's a world of difference between simply passing on information about quality, and taking an active role in promoting quality. In your team meetings, how often is quality mentioned? And what kinds of decisions do you make when there's a conflict between:

  - costs and quality ('The price of this material has gone up. Can we get away with using something cheaper?');
  - output and quality ('We've got customers waiting – if we check it again, we may lose them.').

  Your attitude to quality will be noticed and copied by your team.

- **Taking complete responsibility for the quality of the team's work**

  Your response to this one may have been: 'It depends what you mean'. Certainly the team leader takes responsibility for the team, but that doesn't mean to say that he or she carries the whole burden. As already discussed, quality is everybody's responsibility.

- **Taking steps to provide training and other resources**

  Yes, most would agree that this is part of the team leader's job. Above all, the leader is a facilitator – someone who removes obstacles, and empowers the team to achieve its goals.

- **Encouraging the team to raise standards**

  This is part of setting the right climate for quality. Sometimes remarkable progress can be made, simply by making the occasional supportive remark, especially at times when other pressures might make it easier to let standards fall.

  The way you answered the last three points:

- checking work while in progress;
- checking completed work
- checking just about everything

  will depend on your own team and job. Too much interference can be stifling and discouraging. Too little monitoring may allow standards to fall. Getting the balance right is part of the challenge of management.

Session A

So far in this workbook the word 'standards' has already appeared many times. But we need to be more specific about what it means. Standards are the subject of the next session.

Self-assessment 1

1 Explain what is wrong with the following statements.

   a Quality is another word for 'superior'.

   _____
   _____
   _____

   b The organization's quality experts have the main responsibility for quality.

   _____
   _____

   c The marketing and design of a product is quite separate from its quality aspects.

   _____
   _____

   d Quality is everyone's business, so managers have no special role to play.

   _____
   _____

2 Match each definition on the left with the correct term, taken from the list on the right. Some may be used more than once.

   a Fitness for purpose.

   b The totality of features and characteristics of a product or service that bear on its ability to satisfy stated or implied needs.

   c The degree to which the specification of the product satisfies customers' wants and expectations.

   d The degree to which the product conforms to specifications, when it is transferred to the customer.

   e The operational techniques and activities that are used to fulfil requirements for quality.

   i Process quality
   ii Quality control
   iii Quality
   iv Conformance quality
   v Design quality

Answers to these questions can be found on page 92.

12

Session A

6 Summary

- The needs of customers is the starting point for quality.

- Quality can be summarized as **fitness for purpose**.

- Producers and suppliers need to know their customers' wants, and whether their products are meeting those wants.

- They must then specify the goods and services they intend to produce or supply: that's **design quality**.

- The next stage is to ensure the defined quality standards are adhered to: that's **process (or conformance) quality**, and it's achieved through systems of quality assurance and quality control.

- Quality is the responsibility of **everyone** in the organization.

- Achieving quality depends on management:
  - displaying commitment;
  - setting standards;
  - providing resources;
  - allowing employees to take responsibility for standards;
  - setting a culture for quality.

- Total Quality Management (TQM) involves every member of the organization in a process of continuous improvement with the aim of satisfying the customers' wants and expectations.

Session B Standards

1 Introduction

In Session A we said that getting the quality right means (among other things) setting well-defined standards. But what standards?

There are many kinds of standards relevant to work situations, including:

- standards of recruitment;
- product standards;
- service standards;
- training standards;
- standards in applying techniques;
- standards of honesty;
- standards of behaviour;
- standards of accuracy.

These all have a bearing on quality. However, our main concern in this session is **quality systems standards**, which define the way a supplier is organized to deliver a product. In particular, we will spend a fair amount of time discussing the quality systems standard BS EN ISO 9000.

2 National and international standards

Who sets the standards used by companies? Many will be set internally. But it has long been recognized that companies of all kinds need common standards of quality.

To meet these needs, there are a number of national and international bodies that publish formal standards on all manner of subjects. In this country, the British Standards Institution (BSI) is the primary standards setting body.

If you were to skim through the BSI catalogue, you would find such diverse standards as:

Your organization will probably have copies of the British standards that are relevant to your work.

- specification for power transformers (BS 171);
- specification for concrete roofing tiles and fittings (BS 473, 550);
- protection against spoilage of packages and their contents by micro-organisms, insects, mites and rodents (BS 1133 Section 5);
- test methods for stabilized soils (BS 1924);
- specification of photographic grade potassium bromide (BS 3307);

Session B

- specification for X-ray lead–rubber protective aprons for personal use (BS 3783);
- guide for comparative testing of performance of detergents for hand dishwashing (BS 6584);
- specification for mail payment orders (BS 6601).

On the international front, a good deal of progress has been achieved within the European Union (EU) on agreeing common standards. Since the end of 1992, every member country in the EU has had to recognize the national standards of all the other members.

Many BSI standards are established as being the equivalent of the standards of other international bodies, including the International Standards Organization (ISO).

3 Product standards and quality systems standards

People and companies buying goods and services have always wanted reassurance that their purchases will be of the right quality and reliability.

For you and I, as individual purchasers, the options are restricted. We make our choice on the best information we can get about a product, and the company supplying it (which may be very limited). Then we take the product home, or use the service, and hope it performs as it should and is completely reliable. If we aren't satisfied we may complain to the supplier, and perhaps even ask for an exchange or our money back.

Organizations that buy lots of goods are in a better position – they have greater 'buying power'. They can demand high standards and threaten to take their business elsewhere if they don't get them.

But large purchasers have long recognized that it is not enough to test and inspect the actual goods themselves. The quality of a product or service is a reflection of the way the manufacturer is organized and managed.

There is a clear distinction here between **product standards** and **quality systems standards**:

A product standard tells us about the quality of a product. A quality systems standard tells us about how well the organization is structured to supply that product.

Session B

Activity 6

3 mins

What can large-scale purchasers of goods and services do to ensure that the supplier is well organized to provide goods of the required quality? Jot down one thing they could do.

___

___

There are two basic approaches to this problem of assessing an organization's quality system. The purchasing organization can:

- lay down its own standards and send in its own assessors to ensure these standards are met; or
- insist on the supplier meeting published standards on quality, and on their being assessed against those standards.

A really large purchaser may carry out both of these procedures. For instance, the Ministry of Defence insists on suppliers meeting the NATO standards known as AQAP (Allied Quality Assurance Publications), and will send in MoD inspectors to ensure these standards are adhered to.

In the commercial world, BSI is able to assess an organization's quality system against the standard **BS EN ISO 9000**. Achieving accreditation to this standard is a considerable achievement, and can make a substantial difference to the standing of producers and suppliers in the marketplace. Let's look more closely at this standard now.

4 The quality systems standard BS EN ISO 9000

EXTENSION 2
BS EN ISO 9000 Made Simple, by John Shaw, is a good general guide to this subject.

ISO 9000 is no more than commercial common sense, best business practice and soundly documented controls.[1]

BS EN ISO 9000 was derived from the British Standards specification BS 5750. From now on we'll refer to it as simply ISO 9000.

ISO 9000 is written in an extremely concise way, and has to be read very carefully. It is divided into three parts: ISO 9001, ISO 9002 and ISO 9003.

[1] John Shaw, *BS EN ISO 9000 Made Simple*.

Session B

> An interesting definition is that for product. A product may include service, hardware, processed materials and software, or any combination of these. It may be tangible or intangible; an example of an intangible product is a service to supply information. The term 'product' refers to the goods or service at all stages of the process, including raw materials, sub-assemblies, and the final product.

- If your organization is involved in design, development, production, installation or servicing, it may well have approval to ISO 9001.
- ISO 9002 is applicable if your organization is concerned with production, installation or servicing, where there is no design element. This part of the standard is relevant to most organizations who supply goods or services, and is the most widely adopted standard.
- ISO 9003 is the least used part of the standard, and applies where conformance to specified requirements can be established adequately by inspection and testing. It may be applicable in some divisions of an organization that also has ISO 9001 and ISO 9002 approval.

Each of these parts of ISO 9000 is divided into four sections. Section 1 defines the scope of application. Section 2 lists the documents, including other standards, that are referred to. Section 3 provides the definitions of terms used.

Section 4 of part 1 is a description of the quality system requirements, and is divided into twenty elements, which are briefly summarized below. You are not expected to remember every point, but you should read through the summaries carefully, and complete the activities. The two main reasons for spending time discussing this standard in detail are that it:

- is a good starting point when you want to investigate the subject of standards accreditation for your own organization;
- can help you identify the essential requirements for the achievement of quality, and reinforce the points made in our earlier discussions on the subject.

ISO 9000: 4.1 **Management responsibility**

Every organization is required to have a documented **quality policy**, which must be read and understood by all employees. This can be complied with by providing staff with individual copies of the policy statement, or else displaying the policy in a prominent place.

Organizational responsibilities must be defined, and this usually means creating an **organization chart** (showing job titles rather than names), which shows how the business is structured.

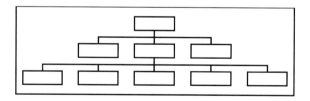

Session B

A key general question to be answered is **what happens when something goes wrong?** – when there is a fault with the product, say, or when a supplier creates a problem. When this happens, heads of departments and other managers are typically expected to:

- identify and record the quality problem;
- take action or suggest ways of preventing any repeat of the problem;
- make sure that proposed solutions are effective;
- keep any non-conforming product under control.

Section 4.1.2.2 of the standard covers the **resources** that are needed for verifying that the quality system is working properly, by means of regular internal audits.

Activity 7

2 mins

We've already talked about resources. Which particular resource would you expect to be key to the successful implementation and checking of a quality system?

_____

_____

> The Quality Assurance Manager is not 'the person responsible for quality'. He or she is the quality standards expert who guides and oversees the organization's quality system.

The most crucial of any organization's resources are its people. And, as we've discussed (but it's important enough to reiterate), all those involved in management, work performance and verification should be **fully trained**.

A key person in any organization implementing ISO 9000 is the person who has the responsibility to ensure that all aspects of the standard are complied with, and who reports to the organization's senior management on the performance of the quality system. In the standard, this person is called the **Management Representative**, but is typically the Quality Assurance Manager or Quality Director.

Management is required to review the system periodically.

Session B

ISO 9000: 4.2   Quality system

Under this element of the standard, the organization must implement and document the policies, procedures and instructions for a quality system. Key to this is a **Quality Manual**, which sets out all the procedures for implementing the system.

Activity 8

Portfolio of evidence A1.1, A1.3 — 15 mins

This Activity is the second of a series of three which together may provide the basis of appropriate evidence for your S/NVQ portfolio. The others are Activities 4 and 9. If you are intending to take this course of action, it might be better to write your answers on separate sheets of paper.

Obtain a copy of your organization's Quality Manual. Such a document will normally exist, whether or not the organization has sought, or intends to seek, accreditation to a quality systems standard.

For this activity you should make a plan for at least one way of improving your team's compliance with this manual. You will probably need to start by reading through the document, and perhaps having parts of it explained to you by someone with detailed knowledge of its contents.

You could jot down some brief notes as you read, using a piece of paper with the headings below.

What does the manual say should be done?	How well does my team comply?	What could we do to improve?

This list does not need to be very detailed or comprehensive. As soon as you think you have an idea for improving your team's performance in respect of the organization's quality requirements, you can get on and set out a fuller plan. You will need to write down:

- what your team does now, in relation to a particular Quality Manual requirement;
- exactly what actions you intend to take, so as to make improvements.

Write your plan on a separate sheet of paper.

20

Session B

ISO 9000: 4.3   Contract review

Every **contract** or order must be checked and documented to ensure that:

- the customer's requirements are defined and understood;
- the organization is capable of meeting those requirements.

ISO 9000: 4.4   Design control

For organizations with **design** capabilities, this element states that formally documented procedures must be established to control and verify the design of a product, in order to ensure customers' requirements can be met.

ISO 9000: 4.5   Document control

This element refers to the control of internal **documentation** relating to the standard, such as: purchase order forms; delivery notes; job cards. The system must ensure that all vital documents are properly approved, and that the current level of issue revision is in use. Changes to documents must be agreed with the Quality Assurance Manager, and obsolete documents withdrawn from the system.

ISO 9000: 4.6   Purchasing

Ideally, the companies that supply goods and services to an ISO 9000-accredited organization should themselves have ISO 9000 accreditation. Failing this, it is important that all suppliers should be assessed, in terms of their quality assurance system. Purchasing documents should state the organization's exact requirements.

ISO 9000: 4.7   Control of customer-supplied product

Goods and services supplied may be inspected on the supplier's premises, by the customer. Whatever system of checks are in place, the purchasing organization must have a high level of confidence that quality levels meet its needs.

Where items are supplied by a customer for incorporation into a product, they must be separately documented, controlled and protected.

ISO 9000: 4.8   Product identification and traceability

This element requires that:

- components, services and other 'bought-in' goods be **traceable** to the company that supplies the product;
- manufactured products must be **traceable** to suppliers of the raw materials, production batches and, if appropriate, to specific operators, shifts or machines.

In practice, this means that all components should be clearly **identified**, and, if necessary, stored separately from one another.

Session B

ISO 9000: 4.9 **Process control**

> Don't forget — these are just brief summaries of the points in the standard. You will need to get more help and advice if you're planning to work towards ISO 9000 accreditation.

The standard states that production, installation or servicing work **processes must be documented and controlled** to prevent error, and to ensure the customer's exact requirements are met.

To control a work process, ISO 9000 says it is necessary to:

- provide work **instructions** to staff so that they are clear about what they are meant to do;
- give them sufficient **training** in order that they are competent at their jobs;
- devise **plans** for production in advance to ensure compliance with customers' requirements;
- provide **equipment** that is fit for purpose and is properly maintained;
- identify **stages of production**, and document these so that product faults can be identified and rectified quickly.

Activity 9

Portfolio of evidence A1.1, A1.3 — 15 mins

This Activity is the third of a series of three which together may provide the basis of appropriate evidence for your S/NVQ portfolio. The others were Activities 4 and 8. If you are intending to take this course of action, it might be better to write your answers on separate sheets of paper.

Part of your job is to control work processes, so as to meet quality needs and maintain work flow. Consider each of the **five** topics, listed in 4.9 above, under this element of ISO 9000, and set out an action plan or proposal for improving performance in at least **one** of these areas.

For example, you might suggest ways in which written instructions given to your team could be less ambiguous, or could provide more useful information. Or you might find ways of improving the planning of production of goods or services, perhaps by improving the co-ordination between your team and another.

Whatever you decide upon, your action plan should be specific, so that it's clear what you intend to do, and when and how you will do it.

Write out your action plan on a separate sheet of paper.

Session B

ISO 9000: 4.10 **Inspection and testing**

Three inspection and testing operations are identified:

- incoming goods
- in-process work
- the final product

Incoming goods need not be 100 per cent inspected or tested, and a company may rely on its suppliers to provide evidence of verification of goods and services. In-process checks must be carried out at each production stage to ensure conformity with requirements. The final product must also be verified. All verification must be fully documented.

ISO 9000: 4.11 **Control of inspection, measuring and test equipment**

To meet the requirements of this element:

- All items of equipment used for inspection, measuring or testing must be recorded in a register, which should also give details about the frequency and nature of the checks and the standards or tolerances which must be adhered to.
- All designated equipment must be calibrated and traceable to national or international standards (for example, weight-measuring equipment would have to be checked against known standard weights which themselves will have been verified by a laboratory).
- Equipment must be stored and handled appropriately to ensure that its accuracy is maintained.

> In Sessions C and D of this workbook, we'll be looking at specific techniques used in inspection and testing.

ISO 9000: 4.12 **Inspection and test status**

In essence, this element requires the organization to set up its inspection and test system so as to be able to answer the following key questions about every product, through all stages of production:

- Has the product been tested?
- Has it passed or failed?

ISO 9000: 4.13 **Control of non-conforming product**

The status of the product must be clearly identified at all stages. Formal written procedures must ensure that a product that does not conform cannot be processed further, or released to the customer.

Session B

Activity 10

If a fault is found with a product (goods or service) so that the organization realizes that the product does not meet the agreed quality specification, what might the organization do with the product? Try to list three possible actions it might take.

___

If a product does not conform to its specification or to the customer's requirements, there are several possible alternative actions, depending on the nature of the product and the specific details. The supplier might:

- take corrective action, such as giving the employees performing the service more training, or replacing a damaged component;
- regrade the product: if it is not suitable for normal sale, it may be appropriate to use it in another way or for a different purpose, an example being that of the 'imperfect goods' you sometimes see for sale in shops at a reduced price;
- scrap the item;
- in the case of goods being found faulty on receipt, return them to the contracted supplier;
- deliver the product to the customer after advising him of its status, and perhaps agreeing a discounted price.

Whatever action is taken, it must be recorded and be consistent with documented procedures.

Session B

Activity 11

Having taken action to deal with a non-conforming product, what, in general terms, should the organization do next?

___

ISO 9000: 4.14 **Corrective and preventative action**

As suggested by the title of this next element, once problems have been discovered it is important to learn from them and to take action to prevent them happening again.

When:

- there is a product fault;
- a supplier does not deliver goods or services as agreed;
- customers complain;
- there are internal problems within the organization;

then the cause must be identified, and steps taken to prevent recurrence.

The standard requires that the problem and the corrective actions are documented, and that the effects of these actions continue to be monitored. The steps are summarized in the figure below.

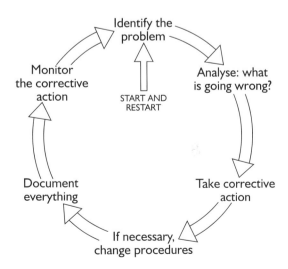

Session B

ISO 9000: 4.15   **Handling, storage, packaging, preservation and delivery**

This element contains the 'housekeeping' provisions of the standard, which relate to:

- protecting the product from damage or deterioration;
- storing the product and the packaging in appropriate conditions;
- controlling the packing process through documented procedures;
- controlling preservation methods such as cold storage;
- delivering the product in sound condition.

ISO 9000: 4.16   **Control of quality records**

This requires quality records to be maintained, and held for an appropriate period. The *actual* period of retention will depend on the products that the organization is producing but typically will be at least three years.

Activity 12

5 mins

What 'quality records' do you think are necessary to manage a quality assurance system? Try to list **four** or **five**. If you can't answer this question, write down the names of any quality documents that you use in your work.

_____
_____
_____
_____
_____
_____

EXTENSION 2
Examples of all these documents are included in the book *BS EN ISO 9000 Made Simple* by John Shaw.

The standard mentions the following quality records:

- Quality audit schedule, quality audit checklist, and quality audit report

  Internal systems checks of the quality system (called audits) must be pre-planned. The schedule lists dates and times of audits; the checklist sets out questions and comments related to audits; the report contains information gathered during an audit.

- Non-conformance reports and registers

  When some non-conformance arises, managers (typically, department heads) must complete a report. The non-conformance might be related to a system problem, a customer complaint, a product fault or some other aspect of quality. The register is used to summarize reports, so that it can be easily seen which department, person or supplier is causing the problem.

Session B

- **Corrective action report**

  The Quality Manager completes this report, should it be necessary to change an operating procedure.

- **Suppliers' non-conformance letter**

  Should a supplier cause a serious problem, a letter is sent, in which a formal warning is issued, threatening the removal of the supplier from the suppliers'/subcontractors' register.

- **Controlled documents issue register**

  This document acts as a control for all other important internal documents.

- **Quality/operational manual amendment control form**

  This form is used to record changes to the Quality Manual or the Operational Procedures Manual.

- **Suppliers'/contractors' register and questionnaire**

  The register records suppliers' details, following the completion of the questionnaire, which requests details of their ISO accreditation status.

- **Training record**

  This is a record of the experience, qualifications and ISO 9000 training of all employees, including senior staff.

- **Calibration register**

  As we discussed earlier, all equipment must be calibrated. This document records what equipment is kept and the date, frequency and type of calibration to be carried out.

- **Customer complaint/acknowledgement letter**

  If there are serious or repeated customer complaints, this advises the customer about the organization's intention to investigate the matter, and what corrective actions will be taken.

ISO 9000: 4.17 **Internal quality audits**

As already mentioned, internal quality audits are required to ensure the quality system is working satisfactorily. This element describes what has to happen, when.

ISO 9000: 4.18 **Training**

To achieve the requirements of the standard, adequate and effective training must be given to all staff including training in quality assurance. As mentioned above, full records must be kept.

Session B

ISO 9000: 4.19 **Servicing**

Many kinds of work involve servicing: maintaining products or overhauling them. The standard requires that servicing is pre-planned and documented, even if it is carried out by a third party.

ISO 9000: 4.20 **Statistical techniques**

Many activities may involve the use of statistics, including: production quality control, equipment performance, market analysis and reliability analysis. Techniques must have a sound statistical basis and be performed according to documented procedures.

We will look at a number of statistical techniques in the next two sessions of the workbook.

Although we have looked in detail at this standard, it is worth remembering that, important as BS EN ISO 9000 is, it is not the only standard that is significant for organizations. The **Investors in People (IIP)** award is often seen as being complementary to ISO 9000. It focuses on the management of people and the contribution that this management makes both to the overall business mission and strategy and to delivering high quality. Recruitment, development of staff and communications are among those aspects covered. Because IIP highlights the **people** in the organization, it is regarded as a way of humanizing the concentration on **systems** in ISO 9000.

Another set of 'people' standards are the National Vocational Qualifications (NVQs). You may be aiming for an NVQ through your study of Super Series workbooks.

Session B

Self-assessment 2

1 For each question listed on the left, select the most appropriate answer from those on the right.

a What is the reason for assessing a supplier's quality system?

b Explain the purpose of one essential element of process control.

c What is the purpose of internal, pre-planned quality audits?

d What is the role of the management representative?

e Why should quality records be retained for a long period?

f What should the organization's approach be to documents related to quality?

g Why should organizations identify inspection and testing stages?

i To review the quality system, and make sure it is working according to the standard.

ii To act as the key person in implementing ISO 9000 in an organization.

iii To carry out checks at each stage, so as to ensure conformity.

iv To provide clear work instructions.

v To make sure that required standards can be maintained.

vi To allow a check to be made of a product's conformity during (and perhaps after) its working life.

vii To keep them under control, and withdraw obsolete ones from the system.

2 Here's another set of questions. Again, for each question listed on the left, select the most appropriate answer from those on the right.

a What is the reason for identifying non-conforming products?

b What is the role of training in the implementation of ISO 9000?

c What is the purpose of calibrating equipment?

d What should the organization's approach be regarding customer orders and contracts.

e What should the organization's approach be when a supplier does not deliver goods or services as agreed?

f What is the function of the Quality Manual?

g Why identify the inspection and test status of a product?

i To ensure that everyone involved is capable of meeting the standard of work required.

ii To ensure standards of measurement are consistent.

iii To ensure that it has been verified and whether or not it conforms.

iv To take action to prevent it happening again.

v To set out all the procedures for implementing the quality system.

vi To make sure they are properly understood, and that the organization is capable of meeting the customer's requirements.

vii To keep them under control, so that appropriate action such as rework can be taken.

Answers to these questions can be found on page 93.

5 Summary

- In this country, the **British Standards Institution (BSI)** is the primary standards setting body.

- A good deal of progress has been achieved within the European Union (EU) on agreeing common standards.

- A **product standard** tells us about the quality of a product. A **quality systems standard** tells us about the way the supplier is organized to deliver that product.

- Achieving accreditation to **ISO 9000** is a considerable achievement and can make a substantial difference to the standing of producers and suppliers in the marketplace.

- ISO 9000 is written in an extremely concise way and has to be read very carefully. It is divided into three parts: ISO 9001, ISO 9002 and ISO 9003.

- A product may include service, hardware, processed materials and software, or any combination of these; it may be tangible or intangible.

- Under ISO 9000:
  - every organization is required to have a documented quality policy which must be read and understood by all employees;
  - a key question to be answered is: what happens when something goes wrong?
  - a key person in any organization implementing ISO 9000 is the management representative – typically the Quality Assurance Manager or Quality Director;
  - the Quality Manual sets out all the procedures for implementing the system;
  - every contract or order must be checked and documented to ensure that the customer's requirements are defined and understood and that the organization is capable of meeting those requirements;
  - for organizations with design capabilities, formally documented procedures must be established to control and verify the design of a product;
  - changes to documents must be agreed with the Quality Assurance Manager and obsolete documents withdrawn from the system;
  - all suppliers should be assessed in terms of their quality assurance system;
  - where items are supplied by a customer for incorporation into a product, they must be separately documented, controlled and protected;
  - all components must be clearly identified and traceable and, if necessary, stored separately from one another;

Session B

- production, installation or servicing work processes must be documented and controlled to prevent error and to ensure the customer's exact requirements are met;
- three inspection and testing operations are identified: incoming goods, in-process work and the final product;
- all designated equipment must be calibrated and traceable to national or international standards;
- the status of the product – whether or not it has been tested and whether or not it conforms – must be clearly identified at all stages;
- non-conforming products must be identified and kept under control;
- once problems have been discovered it is important to learn from them and to take action to prevent them happening again;
- products must be stored, protected, packed and delivered to ensure conformity with customers' requirements;
- quality records must be maintained, and held for an appropriate period;
- internal quality audits are required to ensure the quality system is working satisfactorily;
- adequate and effective training must be given to all staff;
- servicing must be pre-planned and documented;
- statistical techniques must have a sound basis and be performed according to documented procedures.

■ Having made all these points, don't forget that ISO 9000 is not the only standard important to organizations.

Session C Quality control and statistics

1 Introduction

Quality means conformance to requirements, not goodness.[1]

Our focus of interest now turns to process quality, or quality control. The aim of quality control is to determine whether products conform to an agreed specification.

> **EXTENSION 3**
> The book *Statistics* by Frank Owen and Ron Jones covers all the topics in statistics and probability that we will deal with in this workbook. You may want to go further into the subject by taking up this Extension.

This session is a brief introduction to the statistics used in quality control, and what you learn in this session will be applied in Session D. Statistics is concerned with collecting, analysing, interpreting and presenting data. Statistics is important in quality control because quality control involves handling lots of data – mainly in the form of numbers.

One of the easiest things to do with a set of numbers is to work out the average; that's what we'll look at first.

Then we'll go on to discuss ways of measuring the variability or 'spread' of a set of numbers. Two main techniques are used here: the range and the standard deviation.

2 The average or mean

Imagine you are investigating the time spent on the telephone by your sales team. You arrange to have the calls timed over a period of a month, and collect a lot of data. You look at the figures for the first day and find that forty calls have been made. The time taken, in seconds, for the calls is as follows:

193	21	163	73	110	160	143	50	88	49
142	53	205	105	100	136	184	169	87	142
76	93	102	135	107	120	44	166	202	73
204	45	135	85	95	63	155	147	61	169

How can you work out the average length of time for these calls?

---

[1] Philip B. Crosby (1996) *Quality is Still Free*

Session C

The first thing to do is to total all the figures. Then, knowing the total, you can divide by the number of calls to calculate the average.

The total of all the calls = 4650

The number of calls = 40

So the average, or mean, of this data = $\dfrac{4650}{40}$ = 116.25 seconds

The mean or average of a set of values is the total of all the values divided by the number of values in the set.

The mean is sometimes indicated by the symbol $\bar{x}$ (pronounced 'x bar').

Activity 13

Assume a hotel manager keeps records of the guests booking in each month. Guests are asked to fill out a questionnaire asking them what they think of the service, food, room cleanliness and so on. One question asks them to grade the quality of service according to the following grades:

- excellent (A)
- very good (B)
- good (C)
- fairly good (D)
- poor (E)
- very poor (F)

Of those filling out the questionnaire, the following numbers were counted for the answers to this question.

Grade	June	July	Aug	Sept	Oct	Nov	Dec	Totals
A	182	192	176	162	134	208	178	1232
B	148	166	102	98	155	176	188	1033
C	113	78	123	87	67	122	72	662
D	133	108	78	122	151	76	122	790
E	48	66	32	12	27	55	21	261
F	9	10	5	1	17	3	2	47
							Overall total	4025

34

Session C

Now work out the following means.

a Look at the total for grade A. What is the average number who awarded grade A per month?

_____

b Now look at the overall total. What is the average number of questionnaires completed per month?

_____

The answers are as follows.

a The average number who awarded grade A per month is the total number awarding grade A, divided by the number of months

$$= \frac{1232}{7} = 176$$

b The average number of questionnaires completed per month is the overall total, divided by the number of months

$$= \frac{4025}{7} = 575$$

3 The range

Suppose a first line manager in a factory making chinaware is checking the size of plates which have a theoretical size of 250 millimetres. He might pick out ten samples and get the following measurements (in millimetres):

Sample no.	1	2	3	4	5	6	7	8	9	10
Size (mm)	254	244	253	252	240	250	246	264	247	254

Our manager can first total these up and work out the mean.

The total is 2504. Therefore the mean $= \dfrac{2504}{10} = 250.4$ mm.

This is useful, as it would give an indication as to whether the average size is within tolerance. However, the manager would also like to have some measure of the variability of the plate size: by how much do sizes vary? After all, customers won't expect too much variation. One simple measure of variability, or spread of data, is **range**.

The range of a set of values is the largest value minus the smallest.

35

Session C

Activity 14

Here is the list of plate sizes again:

Sample no.	1	2	3	4	5	6	7	8	9	10
Size (mm)	254	244	253	252	240	250	246	264	247	254

In this list, which is the largest?
_____

Which is the smallest?
_____

What is the range for this set of values?
_____

The largest size is sample 8 (264) and the smallest is sample 5 (240).

The range is the difference between the largest and the smallest:

264 − 240 = 24.

This is very useful, because although the mean size of the plates seems to be near to the nominal size of 250 millimetres, the range indicates that there is in fact a lot of variation in size.

The range is a helpful indication of spread, but, as we will see, it has its limitations.

Session C

4 The standard deviation

Standard deviation is one of the most commonly used tools in quality control and is not as difficult to grasp as you might think if you are not used to dealing with figures.

Activity 15

3 mins

Look at the following set of figures:

101, 102, 99, 101, 3, 100, 102, 102, 99, 101.

What is the range?

___

Explain briefly why a measurement of range is not very useful in this context.

___

___

The range of the set of figures given is:

102 − 3 = 99.

However, if we ignore the lowest figure (3) the range becomes:

102 − 99 = 3.

So all the figures are in the range 102 − 99 except for one item which is a long way outside this range.

This example shows the main disadvantage of using the range as a measure of variation: odd single figures can distort the result.

■ A manager of a supermarket was interested in finding out how long customers had to wait in the checkout queues. After measuring times over several days, she found that a customer might expect to wait anything from zero to twelve minutes. However, when she examined the data, she found that only on two occasions out of about 15,000 was the wait as long as twelve minutes; otherwise, it never exceeded eight-and-a-half minutes.

A better method of measuring the spread or variability of data is needed. The one most often employed in quality control is called the **standard deviation**. The standard deviation largely overcomes the disadvantage of being distorted by unusual values that occur only very rarely.

Session C

> To work out the standard deviation on a calculator, you will need one with a square root $\sqrt{}$ function. The square root is a number or quantity that, when multiplied by itself, results in a given number or quantity. So the square root of 4 is 2, and the square root of 9 is 3, that is:
>
> $\sqrt{4}$ = 2 and
> $\sqrt{9}$ = 3.

Working out the standard deviation by hand takes a little more time than working out the range. On a scientific calculator it can be done automatically, simply by keying in each item of data and pressing a key to give the standard deviation function.

If you have access to a computer, standard deviations can be calculated using spreadsheet software as well as on more specialized packages.

Steps for calculating standard deviation

In case you don't have a scientific calculator or computer, the steps for working out the standard deviation by hand are set out below. There are six steps involved. Follow the steps and the example shown. You are not expected to remember the process.

Step	Example
	Data: The lengths of five cut pieces of cloth in centimetres: 123, 128, 113, 127, 125
1 Work out the mean of the set of values.	Total = 616. Mean = $\dfrac{616}{5}$ = 123.2
2 Subtract the mean from each value, to give the 'differences'. (If the mean is greater than the value, the difference will be negative.)	123 − 123.2 = −0.2 128 − 123.2 = 4.8 113 − 123.2 = −10.2 127 − 123.2 = 3.8 125 − 123.2 = 1.8
3 Take each of these differences and square it; that is, multiply it by itself. (Note that the answers are always positive, even if the differences are negative. A minus times a minus is a plus.)	−0.2 × −0.2 = 0.04 4.8 × 4.8 = 23.04 −10.2 × −10.2 = 104.04 3.8 × 3.8 = 14.44 1.8 × 1.8 = 3.24
4 Add up these squares.	0.04 + 23.04 + 104.04 + 14.44 + 3.24 = 144.8
5 Divide this sum by the number of items. The result is called the 'variance'.	$\dfrac{144.8}{5}$ = 28.96
6 Take the square root of the variance; this gives the standard deviation.	$\sigma = \sqrt{28.96}$ = 5.38 cm (approx.)

Session C

So the standard deviation for this example is approximately 5.38 centimetres. This is quite a small figure in relation to the mean, which indicates that the spread or variability of the data is within fairly narrow limits. In other words, the data are tightly clustered, rather than being widely scattered.

The standard deviation is usually denoted by the Greek character σ (pronounced sigma).

Activity 16

8 mins

Work out the standard deviation of the temperature taken from a number of domestic fridges, the data of which are shown below.

Just follow the steps as we did in the example, or else work out the standard deviation directly, using a scientific calculator.

Step	Your calculation
	Data: Five kitchen refrigerator temperature readings in degrees Centigrade: 5°, 7°, 9°, 11°, 13°
1 Work out the mean of the set of values.	
2 Subtract the mean from each value, to give the 'differences'.	
3 Take each of these differences and square it: multiply it by itself.	
4 Add up these squares.	
5 Divide this sum by the number of items. The result is called the 'variance'.	
6 Take the square root of the variance; this gives the standard deviation.	

Session C

The completed table is shown below.

Step	Calculation
	Data: Five kitchen refrigerator temperature readings in degrees Centigrade: 5°, 7°, 9°, 11°, 13°
1 Work out the mean of the set of values.	Total 45. $\bar{x} = \dfrac{45}{5} = 9°C$
2 Subtract the mean from each value, to give the 'differences'.	5 − 9 = −4 7 − 9 = −2 9 − 9 = 0 11 − 9 = 2 13 − 9 = 4
3 Take each of these differences and square it: multiply it by itself.	−4 × −4 = 16 −2 × −2 = 4 0 × 0 = 0 2 × 2 = 4 4 × 4 = 16
4 Add up these squares.	16 + 4 + 0 + 4 + 16 = 40
5 Divide this sum by the number of items. The result is called the 'variance'.	$\dfrac{40}{5} = 8$
6 Take the square root of this variance; this gives the standard deviation.	$\sigma = \sqrt{8} = 2.83°C$ (approx.)

Again, we have a small value for σ relative to the mean, which tells us that the data are tightly clustered.

Don't worry if you found this exercise difficult or tedious. Although it is useful to know how to calculate standard deviations, you would use a scientific calculator or a computer for large quantities of data.

You should note that the standard deviation is in the **same units** as the original data. For instance, the data in the first example were in centimetres, and so was the standard deviation. The second set of data were in degrees centigrade, and we express the standard deviation in the same units.

Session C

5 The distribution of data

So far, we have learned how to work out the mean, the range, and the standard deviation. So, if we have some quality control data, we can find their average value, and measure their spread.

Here's another set of numbers. These are the transit times (in minutes) of lorry-loads of materials travelling between two branches of a company, during one week.

32	27	28	26	31	29	26	31	23	27	26	28	22	23
25	25	30	21	27	26	27	25	24	29	22	20	23	28
28	26	24	24	33	19	25	27	26	25	29	22	27	25
30	29	21	26	24	25	24	28	23	27	25	30	27	28
26	26	24

The quality manager looks at this data and works out the total, the mean and the range.

Activity 17

3 mins

What is the mean and the range of the above data? (To save you adding up the figures, the total is 1534.)

_____

_____

The total = 1534 and the number of values = 59.

So the mean = $\frac{1534}{59}$ = 26.

The lowest and highest numbers are 19 and 33, so the range is 33 − 19 = 14.

Session C

Next, the manager would like to know how frequently each number occurs. The easiest way to do this is to tick them off on a tally chart. He writes down all the possible numbers in the range, then goes through the list, ticking off each number one by one:

		Frequency
19	/	1
20	/	1
21	//	2
22	///	3
23	////	4
24	///// /	6
25	//// ///	8
26	//// ////	9
27	//// ///	8
28	//// /	6
29	////	4
30	///	3
31	//	2
32	/	1
33	/	1

Now the manager can plot these numbers on a **graph**. He draws two lines (called **axes**) perpendicular to one another. On the horizontal axis he marks off a scale showing the times, from one end of the range to the other. On the vertical axis he marks off another scale showing the frequency that each time occurs:

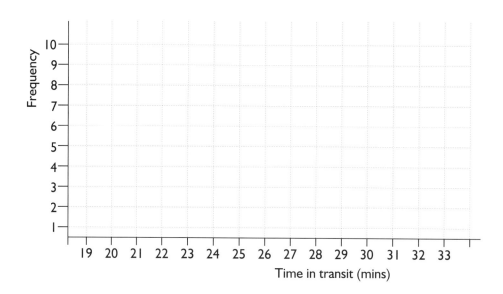

Now our quality manager puts a cross for each value on the horizontal axis, at the point on the vertical axis which corresponds to its frequency. He then joins up the crosses.

Session C

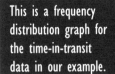

The next figure shows what the graph looks like:

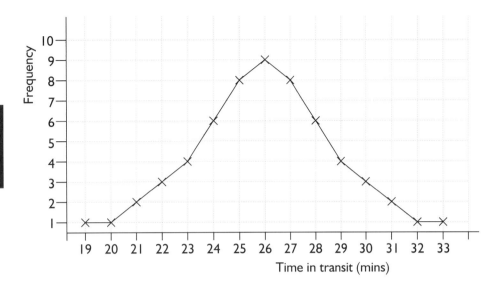

> This is a frequency distribution graph for the time-in-transit data in our example.

The shape of this graph reflects the following facts:

- the curve peaks at 26 minutes, which is also the mean;
- other values occur less frequently, the further away they are from this central value;
- there is a distinct 'bell shape' to the graph.

Of course, the values in this example were carefully chosen so that it would turn out like this. However, it was designed to illustrate some interesting ideas.

If we draw a smooth curve, instead of joining the points with straight lines, we get the general shape shown in the next figure:

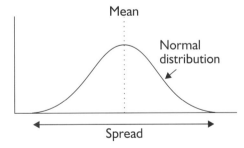

> This is the general shape of the normal distribution.

This is called the **normal distribution**. It is symmetric about its mean. It is bell shaped and the 'fatness' or spread of the bell is measured by the standard deviation of the data. If σ is large, the spread is wide, and if σ is small, the spread is narrow.

The normal distribution is a very special curve, because of the following remarkable fact.

If we take any large population of things or people, and measure some characteristic, the distribution of the data will be normal.

Session C

For example, all of these will be normally distributed:

- the waist measurements of people in any large organization;
- the measured lengths of a large shipment of screws;
- the amount of money spent daily in a shop over a long period;
- the time taken to perform a certain task by individuals in a large group;
- the amount of jam in jam-jars of the same size;
- the number of spelling mistakes occurring per book, in all the books of a publisher;
- the lengths of the incisions made by a surgeon for a particular operation;
- the heights of daffodils on a day in spring;
- the winning numbers in the national lottery over a long period of time.

There are so many possible examples, it would not be difficult to continue this list until we had filled the whole workbook. (And then we could find a normal distribution governing the number of letters per word, and the number of words per page!)

Large populations of all kinds, where there is a tendency to cluster round a mean, display a normal distribution. Going back to our quality manager and his transit times, we noted that the data was approximately normal. Now we can say that, if he were to take a large enough number of samples, the distribution of the data would indeed be normal.

5.1 Varying the mean

As you can see in the diagram, the mean is in the centre of the curve. The spread can be measured by the standard deviation. What happens to the curve if the mean is different?

Activity 18

The general shape of the normal distribution is shown again on the right.

Suppose there are three sets of data, all of which have a distribution the shape of a normal curve, and the same standard deviation, but each has a different mean:

Set A has a mean of 10

Set B has a mean of 20 and

Set C has a mean of 30.

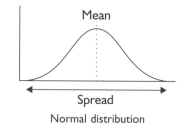

Normal distribution

Session C

On the axes below, sketch the normal curves for these three sets.

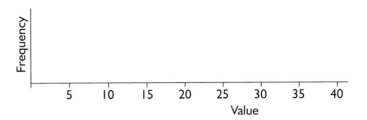

The answer to this activity can be found on page 96.

5.2 Varying the spread

As we've discussed, the standard deviation is a measure of the spread of data. Therefore if one set of data are clustered closely around the mean, they would have a smaller standard deviation than a set of data which vary a long way from the mean.

Activity 19

That being so, what would you expect to happen to the shape of the normal curve as the standard deviation gets bigger? Would it

Get taller and narrower? ☐

Get shorter and wider? ☐

Stay the same? ☐

The answer is that the curve gets shorter and wider as the standard deviation increases. This is what we might expect, because there is a greater 'spread' of data and there aren't so many items of data at the centre.

Session C

The next diagram shows three different normal curves, each with the same mean but with different standard deviations.

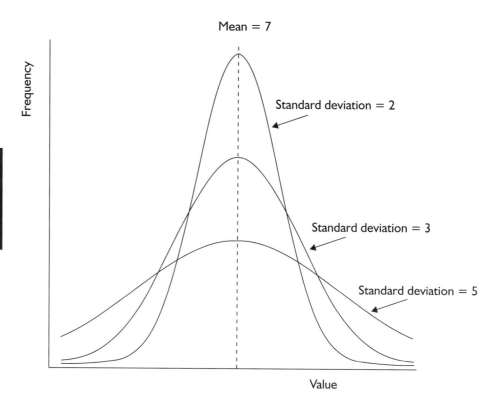

These distribution curves are based on data with the same mean, but different amounts of 'spread'.

In the next diagram, the mean is 20, and the standard deviation is 4. Six vertical lines are drawn showing:

- the mean, plus one standard deviation, marked $+1\sigma$;
- the mean, minus one standard deviation marked -1σ;
- the mean, plus two standard deviations marked $+2\sigma$;
- the mean, minus two standard deviations -2σ;
- the mean, plus three standard deviations $+3\sigma$;
- the mean, minus three standard deviations -3σ.

This is a normal distribution curve of a set of data that have a mean of 20, and a standard deviation of 4 ($\sigma = 4$).

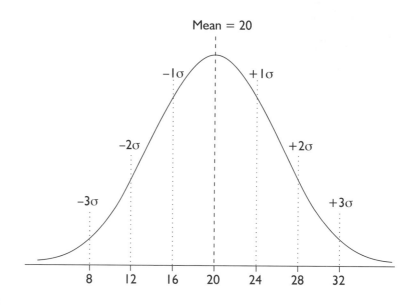

Session C

The standard deviation is 4, so 'mean +1σ' is 20 + 4 = 24. Similarly, you can see that the mean minus two standard deviations is 'mean −2σ' = 20 − 2 × 4 = 12.

Once we get to three standard deviations either side of the mean, nearly the whole of the curve is covered. So almost all of the data lies between the lines '−3σ' and '+3σ'.

The interesting thing is that this doesn't just apply to one particular normal curve – it applies to **all** normal curves.

This means that

once we have worked out the standard deviation for any kind of data which displays the normal curve, we know that nearly all of the data lies between three standard deviations either side of the mean.

To be a little more specific than this, in any normal distribution:

- 68.26 per cent of the data will lie between one standard deviation either side of the mean;
- 95.44 per cent of the data will lie between two standard deviations either side of the mean;
- 99.72 per cent of the data will lie between three standard deviations either side of the mean.

You can see how these add up in the next figure.

As this figure shows, for data which have a normal distribution, nearly all of the data lies between three standard deviations either side of the mean.

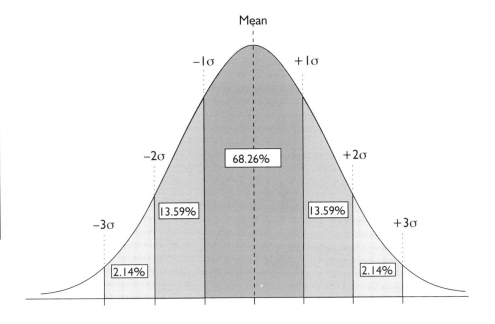

Session C

Activity 20

Looking at the above diagram, what percentage of the data will lie between plus and minus two standard deviations?

___

I hope you agree that a total of 95.44 per cent of the data will lie between plus and minus two standard deviations.

Now try the question in the next activity.

Activity 21

Some data are distributed normally, and have a mean of 50 and a standard deviation of 8. Between what values of data will 99.72 per cent of the data lie?

___

We know that 99.72 per cent of the data will lie between three standard deviations either side of the mean, and that one standard deviation is 8. Therefore:

three standard deviations = 3 × 8 = 24.

So 99.7 per cent of the data will lie between:

50 + 24 = 74 and

50 − 24 = 26.

This is the main point to remember:

99.72 per cent of the data in any normal distribution will lie between three standard deviations either side of the mean.

Session C

Self-assessment 3

1 Find the mean and the range of the following data, taken from a manager's record of the weekly overtime worked by his team:

| 32 | 9 | 72 | 33 | 56 | 18 | 7 | 98 | 93 | 35 |
| 33 | 8 | 10 | 81 | 69 | 84 | 10 | 13 | 59 | 80 |

2 Calculate the standard deviation of the following data set. If you are able to work out the answer on a scientific calculator, use that instead of following the steps listed.

Step	Your calculation
	Data: the typing speed of five typists in words per minute: 80, 57, 72, 48, 63
1 Work out the mean of the set of values.	
2 Subtract the mean from each value, to give the 'differences'.	
3 Take each of these differences and square it: multiply it by itself.	
4 Add up these squares.	
5 Divide this sum by the number of items. The result is called the 'variance'.	
6 Take the square root of the variance; this gives the standard deviation.	

Session C

3 A set of data, which are distributed normally, have a mean of 13 and a standard deviation of 2. Sketch a curve of this distribution on the axes below. Indicate the mean, and mark the position of each of $\bar{x} + 1\sigma$, $\bar{x} + 2\sigma$, $\bar{x} + 3\sigma$, $\bar{x} - 1\sigma$, $\bar{x} - 2\sigma$ and $\bar{x} - 3\sigma$.

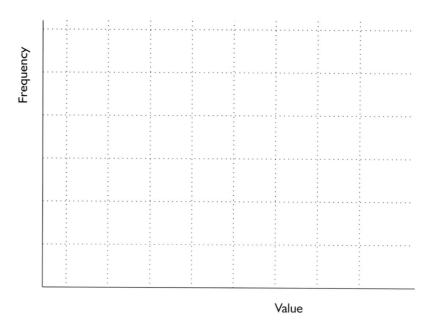

4 Look at the next diagram, which is the normal distribution for a certain set of data. Between which points (from A, B, C, D, E, F, G) will the following percentages of the data lie?

a 68.26 per cent
b 81.85 per cent (there are two possible answers)
c 83.99 per cent (there are two possible answers)

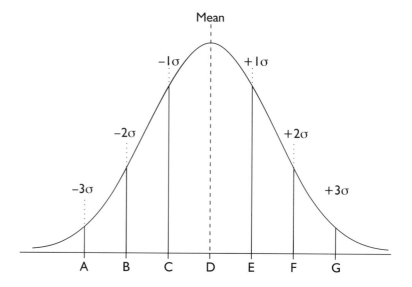

Answers to these questions can be found on page 94.

Session C

6 Summary

- The **mean** or **average** of a set of values is the total of all the values divided by the number of values in the set. It can be called $\bar{x}$ (pronounced 'x bar').

- The **range** of a set of values is the largest value minus the smallest.

- **Graphs** are ways of presenting information in the form of a diagram. A graph illustrates the connection between two sets of data.

- The method of defining the spread or variability of data most often employed in quality control is called the **standard deviation**. Compared with the range, the standard deviation largely overcomes the disadvantage of being distorted by unusual values which occur only very rarely.

- Data of all kinds is **normally distributed**. If we take any large population of things or people, and measure some characteristic, the distribution of the data will be normal.

- Once we have worked out the standard deviation for any kind of data which displays the normal curve, we know that 99.72 per cent of the data in any normal distribution will lie between three standard deviations either side of the mean.

Session D Statistical process control

1 Introduction

> This quotation is taken from the book in Extension 1.

The statistical control of their manufacturing process has been the key to Japan's dominance of world markets ... It is ironic that many of the statistical techniques were pioneered in the West ...[1]

Now that we've covered some basic statistics, we will start to apply it to quality problems.

First we look at one of the most important techniques in quality control: sampling. Sampling means picking out items at random from a large quantity. It is a technique applied whenever it is impractical to check everything, and it is useful in all kinds of work situations. To understand sampling, we need a little probability theory.

We then work through an example of how a sampling plan can be selected to ensure a certain AQL (acceptable quality level).

Next we move to the problem of ensuring that a work process stays within tolerance limits.

2 Sampling

To ensure conformity to an agreed standard, the quality of a product or a component must be inspected or measured.

However, it is impractical to check every item produced or purchased. For one thing, this would be too time-consuming and expensive.

The idea of sampling is that only some of the items are checked. Provided certain defined rules are followed:

the proportion of defects in the sample should give an indication of the proportion of defects in the whole quantity.

---

[1] Dennis F. Kehoe, *The Fundamentals of Quality Management*, p. 135. Chapman & Hall. First edition (1996).

Session D

Sampling is applied not only in all kinds of industry but in other areas of life:

> A **random sample** is a sample chosen in such a way that every sample of the same size has the same chance of being selected.

- When financial auditors check the books of a company, they rarely verify all the invoices and other documents. Instead, they select a certain percentage of them. If errors are found in these samples, the auditors will usually check further. Auditing can be called financial quality control.
- To find out what members of the public think about a certain subject (the outcome of an election), a 'poll' may be undertaken. Here, a random selection of people are asked their opinions. Provided the sample is large enough, and is truly random, the result will be a fairly good indication of the opinions of the public as a whole.
- When a swimming pool supervisor wants to check that the water has the right chemical balance, does not contain too many bacteria, etc., she does not check all the water in the pool, but merely takes a very small sample. In this case, if the water is thoroughly mixed, she can safely assume that the sample is representative of the condition of the rest.
- Many kinds of goods can't be tested without destroying them. For example, to find out how long an electric light bulb will last, the only thing to do is to run it until it fails. So, if you want to work out the average life of bulbs on a production line in a certain period, you can't leave them all on until they fail. But you can run a small random sample of them, and the results will give you a good indication of the average life of the whole batch.

Perhaps you can think of other examples where sampling is used.

3 Probability

Sampling is based on the laws of probability.

> **EXTENSION 3**
> Probability theory is covered in the book *Statistics* by Frank Owen and Ron Jones.

The rest of this section explains some of the theory of probability. It is not difficult but is necessarily brief. I suggest you read it through. You should find it sufficient to get the general idea. The subject is interesting and is applicable to all kinds of work (and leisure!), so if you feel you would like to go into probability further, you might like to take up the suggestion in Extension Three.

If you are playing a board game and you throw a die, there are six faces to the die, so there are six possible outcomes, all equally likely. Therefore there is one chance in six of throwing any particular number – a one, say. We would write this probability as $\frac{1}{6}$.

What if something is sure to happen? What is the probability, say, of getting any number between one and six when we throw the die?

The probability of an event that is certain to happen is 1.0.

On the other hand:

If an event cannot occur, its probability is zero.

Session D

Activity 22

If a die has faces numbered one to six in the normal way, what is the probability of getting a seven in one throw?

The probability of throwing a seven is zero – it can't happen.

If something has a 50/50 chance of happening, we write the probability as ½ or 0.5.

Activity 23

If you toss a (normal) coin, what is the probability of it landing head side up?

The probability of a head is 0.5, because heads and tails have equal chances of occurring.

3.1 The addition rule of probability

Imagine ten playing cards are dealt: six hearts, three spades and one club. The ten cards are well shuffled.

If one card is picked at random:

- the probability of picking a heart is six chances in ten = 6/10 = 0.6
- the probability of picking a spade = 3/10 = 0.3
- the probability of picking a club = 1/10 = 0.1

Notice that:

total probabilities of mutually exclusive events always add up to one.

(By **mutually exclusive** we mean that the events cannot occur at the same time: you cannot pick a card that is a heart at the same time as being a spade, for example.)

55

Session D

What is the probability of picking **either** a spade **or** a club?

To find this, we **add** the individual probabilities:

0.3 + 0.1 = 0.4.

To work out the total probability of one action or another, you add the individual probabilities.

Activity 24

In our example of ten cards, there are six hearts, three spades and one club.

What is the probability of picking **either** a heart **or** a club?

The probability of picking a heart is 0.6 and the probability of picking a club is 0.1. So the probability of picking **either** a heart **or** a club is 0.6 + 0.1 = 0.7.

3.2 The multiplication rule

If we have the same ten cards and pick out a card, we know the probability of picking a spade is 0.3. If a second card is picked (after putting the first card back), what is the probability of picking a spade on **both** occasions?

To get this answer we **multiply** the individual probabilities. So the probability of picking a spade twice in a row = 0.3 × 0.3 = 0.09.

To work out the total probability of one action followed by another, you multiply the individual probabilities.

3.3 Using the rules of probability

Now let's use the rules of probability by looking at an example of sampling to check for defective items in a batch.

Suppose there is a box of ten items and three are defective.

Session D

If one sample item is picked out of the box at random, the probability of the sample being a good item is seven in ten. So we can say the probability of the sample being good = 0.7.

We know that the total probability adds up to one. So the probability of the sample being defective is 1 − 0.7 = 0.3.

What if we take a sample of two? (Assume we replace the first item before drawing the second.) To get the probability of **both** the first sample **and** the second item in the sample being good, we **multiply** the individual probabilities.

The probability of both the first and second items in the sample being good is 0.7 × 0.7 = 0.49. So, when taking a sample of two, the probability of finding **one or more** defects has increased to 1 − 0.49 = 0.51.

Activity 25

3 mins

Following on from this, what is the probability of finding one or more defects when taking a sample of **three**?

_____

_____

Perhaps you can see that the probability of all three samples being **good** is 0.9 × 0.7 × 0.7 × 0.7 = 0.343.

So the probability of finding one or more defects when taking a sample of three has increased to 1 − 0.343 = 0.657.

The greater the size of the sample, the higher the probability:

Sample size	1	2	3	4	5	6	7	8	9	10
Probability of finding at least one defect	0.3	0.51	0.66	0.76	0.83	0.88	0.92	0.94	0.96	0.97

This probability will go on rising as the sample size increases, but will never quite reach 1.0 because, in theory at least, you are never certain to find a defect.

Quite clearly then:

The probability of finding defects rises with a larger sample size.

This is just what we might expect.

Session D

In drawing a sample from a batch, we hope that the number of defects in the sample will exactly match the number of defects in the batch. For any particular sample, this may or may not be so. However, we aim to select a sample size so that, **in the longer term**, with enough batches and samples, there is a high probability that it will be true.

We can imagine a production process that was continuously making boxes of ten items with, say, 1 per cent defective items:

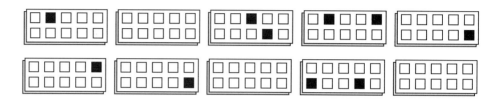

We can take samples repeatedly. Then, provided we choose an appropriate sample size, we can expect, over the longer term, that the proportion of defects found in the samples will turn out to be the same as the actual proportion of defects being produced.

But what is the correct sample size? We'll come back to this subject in a moment.

Meanwhile, as we know that sampling depends on the laws of probability, it follows that we can never predict the results obtained from samples with complete accuracy.

3.4 Sampling or 100 per cent checking?

So sampling doesn't guarantee success. Would it be better, then, to check every item?

Unfortunately, besides being expensive, 100 per cent checking is sometimes not possible.

Activity 26

2 mins

Can you think of an example of goods which could not be checked 100 per cent?

Session D

One example which comes to mind is where the goods have to be destroyed to be tested. How do you ensure that fireworks will produce the desired bangs and flashes? Light every one?

In addition, 100 per cent checking is not foolproof. Inspectors make mistakes, like anyone else. In fact, if the work is repetitive, mistakes occur frequently. Activity 26 may help to demonstrate this.

Activity 27

The paragraph below contains a jumble of letters. Count how many times the letter A appears. Only run through the letters once – if you repeat the exercise you will be carrying out 200 per cent inspection!

AWQRAQWSFAWTASADGDAFFCHASADAFYQATAGADTASARAGFAW
HHAJAHFADREADAEEAWWSFGAJRTDARERAFARWTQAERHNUAKTG
FYTARETTFGAFAGHHGAWDUAYRATAEDETAGDKACAHKHAJAWKAR
UPACOXOSAQHWRRTQAGDHRJELAEADZADASAFRE

Number of letter As: _____

The answer to this activity can be found on page 96.

By concentrating hard you may have got it right, but most people will have miscounted by one or two. If you asked your team to carry out such routine detailed work for a long period of time, you could certainly expect many mistakes to be made.

Before moving on, let's summarize the important points of this section:

- Provided certain defined rules are followed, **the proportion of defects in a sample** should give an indication of the **proportion of defects in the whole quantity**.
- The probability of an event that is certain to happen is 1.0.
- If an event cannot occur, the probability is zero.
- If something has a 50/50 chance of happening, we write the probability as 0.5. Total probabilities of mutually exclusive events always add up to one.
- To work out the total probability of one action or another, you add the individual probabilities.
- To work out the total probability of one action followed by another, you multiply the individual probabilities.
- The more samples that are taken, the greater the chances of finding defects.

Session D

4 Acceptable quality level (AQL)

Imagine you are in charge of inspecting goods coming into a factory, or a department store.

Activity 28

If you choose to sample the goods, can you guarantee to find all defects?   YES   NO

If 100 per cent checking is used, can you guarantee to find all defects?   YES   NO

> **EXTENSION 4**
> If you'd like to study the subject of AQL and other aspects of quality control in greater depth, *A Practical Approach To Quality Control* by R.H. Caplen is a good source of information.

As we have already discussed, no amount of sampling will **guarantee** that all defects are found; neither will checking every item.

Nevertheless, organizations have to control quality, and that often means sampling, because checking every item just isn't feasible. In practice, organizations checking goods have to ask the question:

'What percentage of defects can we tolerate at this point in the operation?'

This may seem a strange question in view of our discussions in the first two sessions of this workbook. However, it is a realistic one. The whole purpose of quality control is to find out the level of quality being reached, so that it can be improved if necessary.

Acceptable quality level (AQL) is the maximum percentage of defects in a sample that can be considered acceptable as a process average.

Of course, the aim of any organization concerned with quality is of course to have **zero** defects, and the term 'acceptable quality level' should not be misunderstood. The question is never: 'How many mistakes are we allowed to make?', because AQL is not intended as a way of letting organizations relax their standards. AQL is only designed to answer the question: 'How well are we doing at this stage?'.

The term 'Acceptable quality level (AQL)' does not mean that it is acceptable to make mistakes or to lower your standards.

Once the AQL is decided upon, the next question is:

'What size sample must we take in order to have a high probability of achieving the AQL?'

60

Session D

> You can find sampling inspection tables in the British Standard BS6001. BS6000 is a guide to the use of BS6001.

It is possible to calculate this for any particular size of batch and AQL, using the laws of probability. However, it is a lot of work and, fortunately, we don't need to, as there are published AQL **sampling plans** or **sampling inspection tables**.

Let's take an example and use a sampling inspection table:

- Items of kitchen equipment are purchased by a mail-order company which intends to re-sell them to its customers. Naturally the company wants to be fairly certain that the items meet the required specification, so it arranges for each incoming batch to be sampled.

  Let's assume in our example that the parts are purchased in batches of 600. Let's also assume that the inspection team is working to an AQL of 1 per cent. This means that if a defect rate of more than one item in 100 – or six in 600 – is found, the batch will be rejected. The question is 'how big should the sample be?'.

The AQL sampling inspection table will show three figures for this size of batch (600) and AQL (1 per cent):

- the sample size – how many samples to check in each batch;
- the number of defects allowed in these samples, in order for the batch to be acceptable;
- the number of defects which, if found in the samples, will result in that batch being rejected.

An excerpt from the published sampling inspection table is shown below.

AQL	0.065%			0.10%			0.15%			0.25%			0.40%			0.65%			1.0%			1.5%			2.5%		
Batch size	n	P	F	n	P	F	n	P	F	n	P	F	n	P	F	n	P	F	n	P	F	n	P	F	n	P	F
281–500	200	0	1	125	0	1	80	0	1	50	0	1	32	0	1	80	1	2	50	1	2	50	2	3	20	1	2
501–1200	200	0	1	125	0	1	80	0	1	50	0	1	125	1	2	80	1	2	80	2	3	80	3	4	32	2	3
1201–3200	200	0	1	125	0	1	80	0	1	200	1	2	125	1	2	125	2	3	125	3	4	125	5	6	50	3	4

Session D

How do we read this table?

- We first find the correct column heading for the AQL we are interested in. For the kitchen equipment, this is the column headed 1.0%.
- Then we look for the correct batch size in the column on the left. In this case it is 501–1200.
- At the point where these columns meet, there are three figures: $n = 80$; $P = 2$; $F = 3$. n is the sample size and P is the maximum number of defects allowed for the batch to pass. If F or more defects are found in the sample, the batch will be failed. So for our example:
  - the sample size required = 80;
  - if up to two defects are found, the batch should be passed, but if three or more defects are found, the batch should be rejected.

Activity 29

Suppose we are working to an AQL of 0.25 per cent and have a batch size of 450:

- How many samples should be taken from each batch? _____
- What is the maximum number of defects allowed in the sample for the batch to be passed? _____
- What number of defects (as a minimum) found in the sample would result in the batch being rejected? _____

See how your answers compare with the following. If you got different figures, check the table again.

For this example:

- 50 samples should be checked in each batch;
- if no defects are found, the batch should be accepted;
- if one or more defects are found, the batch should be rejected.

If this sampling plan is followed there is a high probability that the number of defectives will not exceed the required AQL. Over the longer term, sampling plans like this can be relied upon.

Session D

5 Control limits

Now let's consider another kind of problem in quality control: how to monitor the production of some item so that it continues to be made to specification.

For our example this time, assume that a lathe is being used to make spindles.

The spindles must be manufactured to fairly close dimensions or they won't fit properly in a bearing. However, it would be impossible to make them all precisely the same size – there's bound to be some variation. The designer of the component would set a **tolerance** on the dimensions: the specification would give the dimensions and state that they should be made to these dimensions **plus** or **minus** (±) some amount.

As an example, let's say that the spindle has a nominal diameter of 52 mm. The tolerance is 51.95 to 52.05 mm; another way of saying it is that the spindles should be 52.00 mm ± 0.05 mm.

If the process has just been set up, we may not know for sure whether it is capable of making the spindles within tolerance, so we would need to make this assessment.

Once it has been working well for some time, the process may then drift out of adjustment. The aim of the inspection team would be to try to prevent incorrectly sized spindles being produced. This means detecting when the process is drifting out of adjustment so that it can be corrected **before** it starts producing spindles that are out of tolerance.

Activity 30

2 mins

Imagine you are the line manager responsible for making sure the spindles are inspected and within tolerance. Assuming that it would be too time-consuming and expensive to check every item, what alternative approach to this problem could you take?

____

____

As we've already agreed, if it isn't practical to check every item, it seems sensible to **sample**. As manager, you would no doubt instruct your team to take samples of the output of the lathe and check for diameter size.

Session D

Let's imagine that you have suggested the team take five spindles every hour and measure them. They average each set of five measurements and get the following results:

Sample means (the mean of each sample of five spindles)				
52.01	52.00	52.02	51.99	51.96
51.97	51.96	51.98	51.97	51.96

Activity 31

What is the mean of these figures?

___

The mean of these sample means = $\dfrac{519.82}{10}$ = 51.98 approximately.

How should the results be recorded? We could draw a graph. By plotting the sample means, we would get a graph something like this:

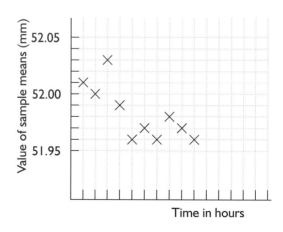

This is a graph of the average of each hourly set of five samples, measured for spindle diameter.

Session D

But we can improve on this diagram. The aim is to produce spindles of 52.00 mm diameter. Also, there is a specified tolerance, so we know that the highest acceptable figure for the diameter is 52.05 mm, and the lowest 51.95 mm. It seems a good idea, then, to draw in lines on the graph for the target diameter, and for the tolerance limits. You can see the result below.

> The tolerance limits provide a visual indication that a process is out of tolerance.

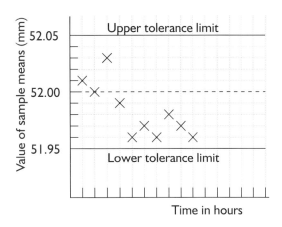

Activity 32

By glancing at the chart above, would you say:

- all the samples are within the tolerance limits?   YES   NO

- there is any drift in one direction or another?   YES   NO

By adding lines for the tolerance limits, I hope you agree that all the samples are in range. It isn't perhaps so easy to detect evidence of drift, but we can see that at the start the readings are near or above the centre line, and later are mainly nearer the lower limit.

5.1 Process capability

The next step is to determine whether the machine is capable of producing spindles within the desired tolerance for a sustained period. This is called the **process capability**. (Although the process has produced some spindles within tolerance, we cannot be sure that this was not pure chance.)

In general, a process is said to be capable if the mean of its sample means, plus and minus three standard deviations, is within the process tolerance limits.

First, we work out the mean and the standard deviation.

Session D

Activity 33

What can we tell about the data by knowing the standard deviation?

_____

_____

One thing we know from our earlier discussions is that we can expect 99.72 per cent of the readings **from any process** to fall within three standard deviations either side of the mean of that process.

The standard deviation of these sample means is 0.021. We already worked out the mean of the sample means as 51.98.

Using these figures:

$$\text{mean} = 51.98 \, \text{mm}; \quad \sigma = 0.021 \, \text{mm}$$

So mean plus three standard deviations = mean + (3 × σ)

$$= 51.98 + (3 \times 0.021) = 52.043 \, \text{mm}$$

And mean minus three standard deviations = mean − (3 × σ)

$$= 51.98 - (3 \times 0.021) = 51.92 \, \text{mm}.$$

This is interesting. The figure of 51.92 is **outside our tolerance limit**. What it tells us is that, even though so far we may have found no samples outside the tolerance band, we can expect that some spindles **will** fall outside this band.

This is the normal distribution superimposed on the tolerance limit chart. (The shaded area shows the part of the distribution outside the tolerance limits.)

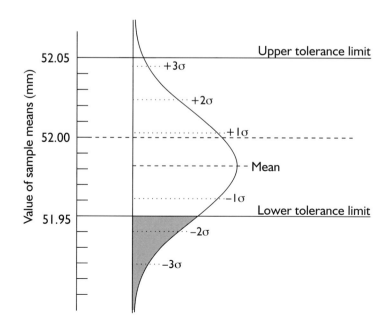

Session D

This is easier to see in the diagram. The normal distribution curve is turned on its side, so that the tolerance band and the spread of the process data are both on the same axis. You can see how the normal curve for the data collected (or rather, that part of the normal curve that is plus and minus three standard deviations from the mean) **overlaps** one of the tolerance limits.

It seems therefore that our process is **not** at present capable – once the mean and standard deviation are calculated, the spread of the data is greater than the spread of the tolerance band. It follows that we would need to make some change to the process to make it capable of producing spindles consistently within tolerance.

Activity 34

Look again at the last diagram. If the data is to be in tolerance, most of the normal distribution curve must be inside the tolerance limits.

How would the data need to be different in order for its normal distribution curve (i.e. three standard deviations either side of the mean) to fit inside the tolerance limits? Pick the correct answer or answers, and then briefly explain your choice.

a The mean of the data would have to be nearer 52.00 mm, with the same spread ☐

b The data would have to be less widely spread, but with the same mean. ☐

c The mean would have to be nearer 52.00 mm **and** the data would have to be less widely spread. ☐

Session D

a Looking at the diagram, let's think first what would happen if the mean of the data were nearer 52.00 mm, but with the data being just as widely spread. If the normal curve were moved upwards, the portion that is plus and minus three standard deviations from the mean still would **not** fit inside the tolerance limits, would it?

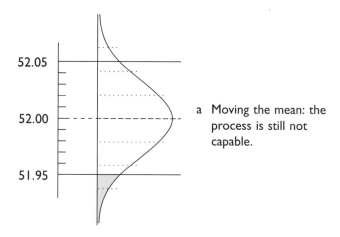

a Moving the mean: the process is still not capable.

b By making the spread of data (and therefore the shape of the normal curve) **a lot** narrower, but **without** changing the mean, the portion that is plus and minus three standard deviations from the mean **could** be made to fit inside the limits.

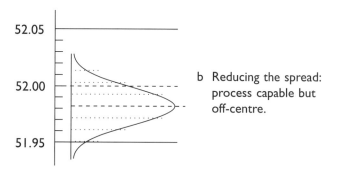

b Reducing the spread: process capable but off-centre.

c If the spread was **slightly** narrower, **and** the mean was nearer 52.00 mm, again it **could** be made to fit inside the tolerance limits.

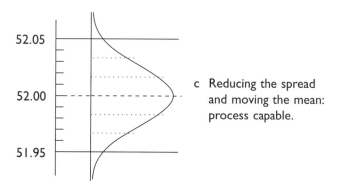

c Reducing the spread and moving the mean: process capable.

So answers b and c are in fact both correct. It's simply a decision of whether 99.72 per cent of the normal distribution will fit inside the 'window' of the tolerance limits.

Session D

We can summarize by repeating the point made earlier:

In general, a process is said to be capable if the mean of its sample means, plus and minus three standard deviations, is within the process tolerance limits.

There is a general formula for process capability, assuming there is an upper and lower tolerance limit. It is:

$$\text{process capability } (C_p) = \frac{\text{total specification tolerance}}{\text{total effective range}}$$

Here, C_p is an index of process capability. If C_p is less than one, the process is **not** capable. If C_p is greater than one, the process **is** capable. The greater the value of C_p above one, the greater the capability.

The **total specification tolerance** is the 'intended range' or 'design range'.

The **total effective range** is the 'actual range' = six standard deviations.

So that:

$$C_p = \frac{T_U - T_L}{6\sigma}$$

where T_U and T_L are the upper and lower specification limits, and σ is the standard deviation for the data.

This follows from what we have already discussed. It is important, of course, that the data is centred.

The following diagram illustrates three possible situations.

a The process is capable.
b The process is not capable.
c The process would be capable if it were centred.

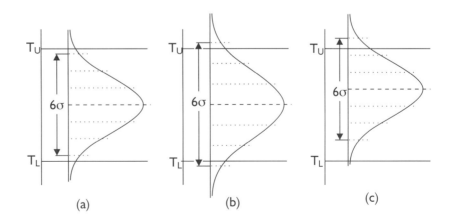

(a) (b) (c)

Session D

Activity 35

Thinking now about the actual work process, what possible reasons might there be for a process not to be capable?

_____
_____
_____

There could be lots of possible reasons. Often, it is the first line manager who has to make the decision about what is causing the problem, and to try to put it right.

The temptation is to ignore what the statistics are telling you. There is usually a lot of pressure on keeping a process going in spite of problems. Managers have even been known to alter the results and pretend the process is still within tolerance, because they know that is the easier course of action.

But work processes go out of tolerance for a variety of reasons, many of them easily rectified. For instance it may be that:

- the lighting is poor;
- a member of staff is inexperienced or needs further training;
- equipment requires adjustment;
- a part on a machine is worn and needs replacing;
- work material is defective;
- there is a problem with the design;
- closer monitoring of results is needed.

Activity 36

Every September, a typical college might process around 15 000 enrolment forms, entering the data into computers. Inevitably errors are made, and these have to be found and corrected.

In one particular college, the managers aimed to reduce the overall error rate as much as possible through a series of measures (extra training, improvement of working conditions, and so on). It set a new initial target of ten errors per person per day.

Once the extra measures had been taken, and the 'process' had settled down, the daily error rate was recorded. The work of five people a day were checked thoroughly as a sample, and the following results were found:

70

Session D

Sample means of number of errors per day (the mean of each sample of five people)				
8	11	9	4	2
7	13	8	2	6

a Work out the mean of sample means and the standard deviation.
b Decide whether this work process is capable. (Hint: there is no 'lower tolerance limit'.)

The answer to this activity can be found on page 96.

5.2 Setting control limits

Assessing process capability is a useful way of determining whether the process is likely to produce output which is consistently within tolerance.

However, once the process has been set up, and is working normally, it may still drift out of tolerance at some point. It would be useful to have an early warning of this, and the procedure below is one way of monitoring the situation.

In Activity 33 we looked at data from spindles produced by a lathe. Here are some different sample readings from the same lathe. Again, these are 'sample means' – the team has taken five samples at a time, averaged the result and then plotted the point.

Sample means (the mean of each set of five samples)					
52.03	52.01	52.02	52.03	52.02	52.03
52.03	52.02	52.03	52.01	52.02	52.02
52.01	52.00	51.99	51.98	51.98	51.98
51.99	51.98	51.98	51.97	51.96	51.96

71

Session D

This time the sample ranges are also calculated – that is, the difference between the highest and the lowest value in the sample – for each set of five samples. These are:

Sample range (the range of each set of five samples)					
0.013	0.010	0.010	0.004	0.003	0.022
0.022	0.021	0.021	0.032	0.023	0.003
0.016	0.012	0.023	0.013	0.018	0.042
0.020	0.005	0.012	0.002	0.020	0.011

Again, it's difficult to make sense of these figures until we plot them on a chart. First, the sample means are plotted, as before:

This is a graph of the average of each hourly set of five samples for our new set of data, measured for spindle diameter.

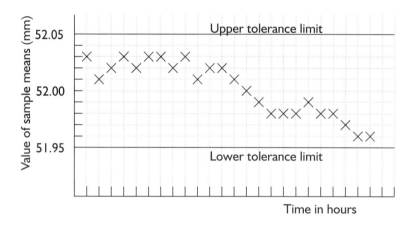

Here there seems to be definite evidence of a drift, even though again all the samples are within limits. This drift towards the lower end of the band suggests that the lathe may need readjustment.

(We could also plot the sample ranges but it wouldn't tell us very much at this stage. We will need to use them though, as you will see shortly.)

When should the operator be told to readjust the machine?

■ When the samples actually go outside the tolerance limit?

This may be too late. Every item outside tolerance is scrap. It is possible to pick good samples even when many defective items are being produced, so the delay may result in a lot of rework.

■ Well before the samples go outside the limit?

Yes – but when exactly?

Session D

One way to solve this problem is to set **control limits** inside the tolerance limits. If we did this we would get a warning that something was wrong before the tolerance limit was reached – which is our aim.

The only question now is **where** exactly to set the control limits.

In fact, the control limits are calculated by averaging the ranges of the samples.

The steps are as follows:

■ First of all we work out the mean range. This is the average of all the sample ranges.

As we discussed earlier, the range of a single sample will only give us a rough indication of its variability. However, the mean range gives a good indication of variability. In fact, there is an exact relation between the true mean range and the standard deviation.

In our example, the mean range works out to 0.01575.

■ Next we multiply this mean range by a constant, according to the sample size. This constant is given in the following table:[2]

Sample size	2	3	4	5	6	7	8
Constant	1.51	1.16	1.02	0.95	0.90	0.87	0.84

Our sample size was 5, so we use 0.95 as a constant. Thus we have:

0.01575 × 0.95 = 0.015 (approx.)

On the average chart, we draw the two control limits so that they are each this distance (Constant × Mean range) inside the tolerance limits.

The **upper control limit** works out to 52.05 − 0.015 = 52.035

The **lower control limit** works out to 51.95 + 0.015 = 51.965.

---

[2] Table quoted in *A Practical Approach to Quality Control*, p. 340. Business Books Ltd.

Session D

The new limits are then drawn in on the chart, as shown in the next figure. The idea is that when a control limit is breached, it is treated as an early indication that the corresponding tolerance limit may be exceeded.

> We have put control limits on our graph, calculated using the range of the data.

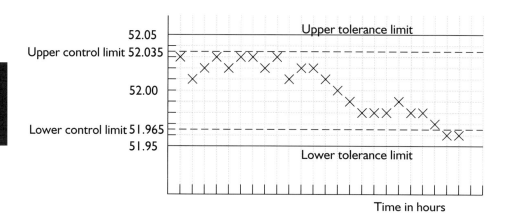

Activity 37

Draw a ring around those samples in the chart above which are outside the control limits.

___

You should have ringed the last two sets of samples at the bottom right-hand corner of the chart. It is at this point that the lathe should be readjusted.

To summarize:

- The problem is one that is frequently encountered in quality control: how to monitor work processes so as to:
  - make sure that they are capable of producing results that lie within agreed tolerance limits, and
  - detect when the process is drifting out of tolerance.

- Sampling is used, and the sample means are plotted on a control chart. The tolerance limits are drawn on the chart.
- Process capability – determined by whether the mean of its sample means, plus and minus three standard deviations, is within the process tolerance limits – is a useful measure of how well the process is likely to produce output consistently within tolerance.
- To give an early indication of a process drifting out of control, control limits are drawn within the tolerance limits.

Session D

6 Applying the techniques

The techniques you have learned in this session of the workbook are used in **statistical process control (SPC)**.

6.1 Statistical process control

Statistical process control is the general term applied to the use of statistics to analyse data from processes (or process outputs) as an aid to solving quality problems.

We used SPC in the last section, when we determined control limits for a machining process.

The most common application of SPC is in manufacturing processes, but it has also been successfully applied to such diverse problems as keeping telephone costs within acceptable limits, and cutting down accident rates. In fact, SPC can usually be applied where:

- there are many individual events taking place, and
- it is desired to set and maintain control limits.

Some possible applications of SPC are as follows:

- A parcel delivery service company gives a guarantee to its customers on delivery times: 'If we're late with your parcel, you get double your money back!' The company knows that a lot of organization goes into getting a parcel from source to destination. It needs to be sure that its work processes are capable, and that any drifting towards an 'out of limits' situation is detected well before it happens.

- A supermarket selling perishable goods works to a system of 'best before' and 'display until/ use by' dates. If goods stay on the shelves too long, they will be wasted. A method of quality control using SPC aims to keep the amount of waste below a certain level. Records of the numbers of out-of-date items for each product line are kept. Used together with sales figures, these give store managers an early indication of changes in customer preferences, as well as helping to keep waste and costs down.

- A bank uses operators to key in vast amounts of data. A certain level of error rate is inevitable, but the aim always has to be to keep the number of keying errors to a minimum. Certain types of critical data are always double-checked, and the rest is sampled. Running records are kept of each operator's performance.

Session D

- A garden plant supplier sells large numbers of plants to garden centres and DIY stores. Quality criteria for each type of plant are defined, including height, number of flowers, number of leaves, and so on. Each batch is sampled, to check that the defined criteria are being met. SPC helps to give an early indication of any fall in quality.

Of course, SPC is simply a tool to help people make the right decisions. SPC won't tell you **what** to do to improve performance. But it will indicate **whether** you need to take action, and **whether** or not those actions have had the desired effects.

SPC is used in answering the following questions about a process:

- Is the process operating in a stable and predictable way?
- By how much can we expect the process to vary?
- Knowing these answers, what targets or control limits could be set?

Suppose a work operation is varying, so that some of the results are outside acceptable limits.

- The first thing you would want to know is how much variation there is in the process before anything is changed.

Here is where statistical analysis of the process data will help.

- Then you would need to decide on the appropriate action: perhaps changing a setting, replacing a machine, or giving staff more training.

Statistics won't tell you what to do – only that some action needs to be taken. Here is where your experience and knowledge of the process come into play in devising an appropriate action.

- Once you have taken this action, you want to know how successful it has been.

Here again, statistics should be of great help.

Let's look at the steps typically used in statistical process control. The relevance of the techniques you have learned in this workbook will then become more obvious.

Session D

Step	Activity	Comment	Techniques typically used
1	Data from the process in question is collected and displayed.	The data must be accurate and should reflect the process performance. Decisions which have to be taken include what size of sample to collect, and how frequently to take samples.	**Sampling** **Probability theory** **Graphs**
2	This data is plotted on a control chart.	This is what we learned to do in the last section. The range, as well as the mean, may be plotted.	**Range** **Mean** **Control charts**
3	Once the process has settled down, its process capability is determined.	If you recall, capability is the ability of a work process to produce output within a desired tolerance for a sustained period.	**Process capability**
4	Control limits are calculated.	There may be pre-defined specification limits, as in our examples. In other processes, there may be no imposed limits: the aim is then to control the process to limits that are as tight as possible.	**Control limits**

6.2 Using statistics to help solve problems

There are many applications for the quality techniques we have looked at in this workbook. When approaching problems involving these techniques, you may find the following hints useful:

- Make sure the data you collect reflects the performance of the process.
- Use charts and diagrams to display the data, so that it can be understood more easily.
- Be patient and careful – don't jump to conclusions.
- Use statistics to help you determine whether the standard you want to achieve is really achievable.
- Use statistics to help analyse results.
- Make only one change at a time – and then use statistics to tell you what difference the change has made to your problem.
- Keep on learning more about statistics, so that you have lots of statistical tools which you can use at will.

Session D

Self-assessment 4

1. A team member is sampling boxes of compact discs. Each box contains twenty discs. A particular box contains the following faulty discs: one scratched disc, two over-size discs and three discs with incorrect labels. The team member picks out one disc at random.

   a What is the probability of selecting a disc with an incorrect label?
   b What is the probability of selecting a disc with a fault of any kind?

   The box then passes to a second team member who picks out another disc (the first disc was replaced, whether or not it was faulty).

   c What is the probability of either person finding a faulty disc?
   d How can the team change the system so that they become certain to find all the faults?

2. A DIY chain samples boxes of tiles for defects. If there are 480 tiles in a box, and they work to an AQL of 0.4 per cent, how many samples should be taken from each box? What is the maximum number of defects that should be allowed in the sample for the box to be accepted? What is the minimum number of defects that should result in the box being rejected?

   Use the following sampling inspection table in your calculation.

AQL	0.065%			0.10%			0.15%			0.25%			0.40%			0.65%			1.0%			1.5%			2.5%		
Batch size	n	P	F	n	P	F	n	P	F	n	P	F	n	P	F	n	P	F	n	P	F	n	P	F	n	P	F
281–500	200	0	1	125	0	1	80	0	1	50	0	1	32	0	1	80	1	2	50	1	2	50	2	3	20	1	2
501–1200	200	0	1	125	0	1	80	0	1	50	0	1	125	1	2	80	1	2	80	2	3	80	3	4	32	2	3
1201–3200	200	0	1	125	0	1	80	0	1	200	1	2	125	1	2	125	2	3	125	3	4	125	5	6	50	3	4

Session D

3 A pizza restaurant aims to deliver an order, within a five-mile radius, in under fifteen minutes. A sample of five measurements are taken each day of the length of time taken for a delivery, and the following results are found for a ten-day period:

Sample means of delivery time in minutes (the mean of each sample of five daily timings)					
13.6	12.4	11.8	14.2	13.2
12.8	11.6	14.8	12.6	13.0

a Work out the mean of sample means and the standard deviation.
b Sketch a suitable graph of these values, showing control limits. (Hint: there is no 'lower tolerance limit', and the values are all above 10, so the vertical scale could be marked from 10 to 15.)
c Decide whether this work process is capable.

Answers to these questions can be found on page 95.

7 Summary

- The idea of sampling is that only some items are checked. Provided certain defined rules are followed, the proportion of defects in the sample should give an indication of the proportion of defects in the whole quantity. The more samples that are taken, the greater the chances of finding defects.

- The probability of an event that is certain to happen is 1.0. If an event cannot occur, the probability is zero. If something has a 50/50 chance of happening, the probability is 0.5. The total probability of mutually exclusive events always adds up to one.

- To work out the total probability of one action or another, you add the individual probabilities. To work out the total probability of one action followed by another, you multiply the individual probabilities.

- Acceptable quality level (AQL) is the maximum percentage of defects in a sample that can be considered acceptable as a process average. It does **not** mean that it is acceptable to make mistakes.

- A problem frequently encountered in quality control is the monitoring of work processes to make sure that results are within tolerance, and to detect when they are drifting out of tolerance.

- Sampling is used, and the sample means are plotted on a control chart. The tolerance limits are drawn on the chart.

- Process capability – the ability of a work process to produce output within a desired tolerance for a sustained period – can be determined by calculating the mean of its sample means, plus and minus three standard deviations, and finding whether this is within the process tolerance limits.

- To give an early indication of a process drifting out of control, control limits are drawn within the tolerance limits.

- Statistical process control (SPC) is the general term applied to the use of statistics to analyse data from processes or process outputs, as an aid in solving quality problems.

Performance checks

1 Quick quiz

Jot down the answers to the following questions on *Achieving Quality*.

Question 1 Complete the sentence: 'The starting point for quality is ....'

Question 2 What's the difference between a **producer** and a **supplier**?

Question 3 What's the difference between **design quality** and the **quality of conformance**?

Question 4 Which people in the organization are involved in quality?

Question 5 Briefly, how can a manager help to 'set a culture for quality'?

Question 6 Briefly, what is meant by 'continuous improvement'?

Question 7 Given the choice between two organizations offering identical products, what greater assurances do you have by buying from the supplier with ISO 9000 accreditation?

Performance checks

Question 8 What is contained in a quality manual?

Question 9 Why is it important that raw materials, services and other 'bought-in' goods be traceable to the company that supplies the product?

Question 10 Under ISO 9000, is sampling of products for conformity allowed?

Question 11 'If we take any large population of things or people, and measure some characteristic, the distribution of the data will be normal.' Briefly explain what this sentence means.

Question 12 What is the probability of an event (a) that is certain to happen; and (b) that cannot happen?

Question 13 Explain the meaning of the term 'acceptable quality level (AQL)'.

Question 14 What is meant by the term 'process capability'?

Question 15 What's the difference between a **tolerance limit** and a **control limit**?

Answers to these questions can be found on page 97.

Performance checks

2 Workbook assessment

Read the following case incident and then deal with the instruction that follows, writing your answers on a separate sheet of paper.

- Harfleet Plastics is a small company in the business of manufacturing goods for domestic equipment. Its Managing Director, Alan Pursloe, would like the company to receive ISO 9000 accreditation and calls you in as a consultant.

  After a preliminary investigation you discover the following facts:

  a In order to 'stay competitive', the company buys raw materials at the cheapest price, regardless of who the supplier is.
  b Although most customers are satisfied with Harfleet's quality, there have been a few complaints, which the company has responded to in a number of ways. As Alan Pursloe says, 'You have to be careful – some people complain about any little thing. We're delivering products that have to be made quickly and cheaply. We try not to cut corners but people get what they pay for. There's no time to inspect everything and sometimes a machine will drift out of tolerance which you can't do much about until it happens.'
  c There are two points of inspection: materials are sampled when they are received, and the final product is sampled before shipment.
  d The company has a small group of design engineers and they sometimes get another company to design items for them. Alternatively, the customer will specify a design. However, drawings and specifications are not always evident on the production floor; the reason for this is that 'people generally know what they're doing – they can ask the engineers if they're in any doubt'.
  e Some of the test equipment seems to be old and of dubious reliability.
  f The Production Manager, Sertan Lescott, also doubles as the Quality Assurance Manager. When you attend a management meeting you observe a discussion about marketing, during which the Marketing Manager, Lesley Ackerman, is heard to remark: 'Don't bring me into discussions about quality – I've enough to worry about. You stick to your job and I'll stick to mine.'
  g When you ask about training, you are told: 'Many jobs are unskilled and people doing these don't need much training so they largely learn on the job. Of course, more skilled people have their own qualifications and training, mostly gained before they joined us. But we don't concern ourselves too much about certificates, provided the job gets done efficiently.'

Write a report to the Managing Director, describing your recommendations. What should your report contain? Make out a list of specific points that you can use to develop for a more detailed report. (Try to include points based on what you have learned from all four sessions of this workbook.) You do not need to write more than a page or so.

Performance checks

 ## 3 Work-based assignment

The time guide for this assignment gives you an approximate idea of how long it is likely to take you to write up your findings. You will find you need to spend some additional time gathering information, perhaps talking to colleagues, and thinking about the assignment. The result of your efforts should be presented on separate sheets of paper.

Your written response to this assignment may provide the basis of appropriate evidence for your S/NVQ portfolio.

The assignment is designed to help you to demonstrate your personal competence in:

- analysing and conceptualizing;
- building teams;
- focusing on results;
- thinking and taking decisions;
- striving for excellence.

What you have to do

Choose **either** A **or** B below.

A   Having learned something of statistical process control, you may feel that it can be usefully applied to work processes carried out by your team. If so, investigate the matter further, preferably after discussing it with someone who is experienced in using these techniques, such as the Quality Assurance Manager.

Your aim is to draw up a preliminary plan setting out your ideas. You will need to answer these questions:

a What process(es) could benefit from the application of SPC?
b Which specific techniques will be employed?
c In broad detail, how will they be applied?
d What further data or information will need to be collected?
e What do you expect to be learned as a result of applying these techniques?
f What will be the benefits to the organization?

Write down your plan in the form of a report, or a memo to your manager.

Performance checks

B Select one of the following areas of quality management and draw up a brief report which explains how your team can contribute more effectively to the organization's quality system or procedures in this respect:

- contract review;
- design control;
- document or data control;
- purchasing;
- product identification or traceability;
- process control;
- inspection and testing;
- inspection,
- test or measuring equipment;
- control of non-conforming product;
- corrective and preventative action;
- storage or packaging;
- training.

Reflect and review

1 Reflect and review

Now that you have completed your work on *Achieving Quality*, let us review our workbook objectives.

- When you have completed this workbook you will be better able to explain the meaning and purpose of quality.

We began by discussing what quality is and what it isn't. The organization's starting point for quality is always the needs of its customers, and quality can be summarized as 'fitness for purpose'. Quality does not mean offering gold-plated luxury goods or sophisticated services to people who have no need for them.

Therefore, producers and suppliers must:

- know what their customers need and want;
- find ways of designing products to meet those needs and wants;
- define quality standards;
- ensure that these standards are adhered to.

- Are you convinced that your team has a good understanding of quality's meaning and purpose? _____

- If not, what training or instruction would be useful to them, in this respect?

_____
_____
_____
_____

Our next objective was:

- When you have completed this workbook you will be better able to describe some sound approaches to quality management.

As we have discussed, effective quality management includes:

- displaying commitment;
- setting standards;
- providing resources;
- allowing employees to take responsibility for standards;
- setting a culture for quality.

Reflect and review

All this is very easy to say, but requires a great deal of planning and effort to put it into practice.

- How might you display more commitment to quality? Write down one thing you could do.

  _____
  _____

- What further resource might you provide which would make it easier for your team to realize their quality goals?

  _____
  _____

The third objective was:

■ When you have completed this workbook you will be better able to summarize the contents and purpose of ISO 9000.

Session B contained a broad summary of the contents and purpose of this standard, so we won't list them all again. As you will have noticed, ISO 9000 imposes a good many demands on organizations and, as you might expect, accreditation to the standard is not easily achieved.

- If you need to find out more about ISO 9000, how will you set about doing so?

  _____
  _____
  _____

- How could you make your team better informed about this standard?

  _____
  _____
  _____

The next objective was:

■ When you have completed this workbook you will be better able to carry out simple statistical and probability calculations related to quality control.

In statistics, we have looked at the mean, the range and the standard deviation. Our brief look at probability was specifically geared towards sampling techniques. If you are not mathematically inclined you may have struggled with these sections. If so, you can decide whether you can afford to forget all about them or struggle on and try to learn more. The book listed under Extension Three may be worth looking at, or you could perhaps find a convenient evening class.

Reflect and review

- What further action will you take to develop your skills in statistics?

  _____
  _____
  _____

The final objective was:

- When you have completed this workbook you will be better able to recognize how the techniques of statistical process control can be usefully applied to work processes.

We have seen that SPC is not confined to manufacturing industry. All kinds of producers and suppliers can benefit from these techniques.

- How could you benefit from SPC?

  _____
  _____
  _____

- What could you do to learn more about the subject?

  _____
  _____
  _____

2 Action plan

Use this plan to further develop for yourself a course of action you want to take. Make a note in the left-hand column of the issues or problems you want to tackle, and then decide what you intend to do, and make a note in Column 2.

The resources you need might include time, materials, information or money. You may need to negotiate for some of them, but they could be something easily acquired, like half an hour of somebody's time, or a chapter of a book. Put whatever you need in Column 3. No plan means anything without a timescale, so put a realistic target completion date in Column 4.

Finally, describe the outcome you want to achieve as a result of this plan, whether it is for your own benefit or advancement, or a more efficient way of doing things.

Desired outcomes				
1 Issues	2 Action	3 Resources	4 Target completion	Actual outcomes

Reflect and review

3 Extensions

Extension 1 Book   *The Fundamentals of Quality Management*
Author   Dennis F. Kehoe
Publisher   Chapman & Hall, first edition (1996)

This book covers most aspects of quality and would be useful for reference.

Extension 2 Book   *BS EN ISO 9000 Made Simple: A practical guide to interpreting and implementing the new international standard*
Author   John Shaw
Publisher   Management Books 2000 Ltd., first edition (1995)

You may find this book useful if you are likely to be involved in getting your organization up to ISO 9000 standard.

Extension 3 Book   *Statistics*
Author   Frank Owen and Ron Jones
Publisher   Pitman Publishing, fourth edition (1994)

You may find this useful as a general introduction to statistics, including the quality control aspects. However, there are plenty of other books on the same subject.

Extension 4 Book   *A Practical Approach to Quality Control*
Author   R.H. Caplen
Publisher   Business Books (Hutchinson), fifth edition (1988; reprinted 1996)

This book deals with the technical aspects of quality, covering the City and Guilds syllabus on the subject.

These extensions can be taken up via your NEBS Management Centre. They will either have them or will arrange that you have access to them. However, it may be more convenient to check out the materials with your personnel or training people at work – they may well give you access. There are other good reasons for approaching your own people; for example, they will become aware of your interest and you can involve them in your development.

Reflect and review

4 Answers to self-assessment questions

Self-assessment 1 on page 12

1 a Quality is another word for 'superior'.

Quality is not another word for 'superior'. On the contrary, quality means producing and supplying products that are fit for their purpose, and meet the customer's needs.

b The organization's quality experts have the main responsibility for quality.

Everyone in an organization shares the responsibility for quality.

c The marketing and design of a product is quite separate from its quality aspects.

All aspects of a product affect its quality, including marketing and design.

d Quality is everyone's business, so managers have no special role to play.

Quality **is** everyone's business, but managers have an important responsibility to lead the way in quality, as in other things.

2

a	Fitness for purpose.	iii Quality
b	The totality of features and characteristics of a product or service that bear on its ability to satisfy stated or implied needs.	iii Quality
c	The degree to which the specification of the product satisfies customers' wants and expectations.	v Design quality
d	The degree to which the product conforms to specifications, when it is transferred to the customer.	iv Conformance quality or i Process quality
e	The operational techniques and activities that are used to fulfil requirements for quality.	ii Quality control

Reflect and review

Self assessment 2 on page 29

1.
   a. What is the reason for assessing a supplier's quality system?

   b. Explain the purpose of one essential element of process control.

   c. What is the purpose of internal, pre-planned quality audits?

   d. What is the role of the Management Representative?

   e. Why should quality records be retained for a long period?

   f. What should the organization's approach be to documents related to quality?

   g. Why should organizations identify inspection and testing stages?

   v. To make sure that required standards can be maintained.

   iv. To provide clear work instructions.

   i. To review the quality system, and make sure it is working according to the standard.

   ii. To act as the key person in implementing ISO 9000 in an organization.

   vi. To allow a check to be made of a product's conformity during (and perhaps after) its working life.

   vii. To keep them under control, and withdraw obsolete ones from the system.

   iii. To carry out checks at each stage, so as to ensure conformity.

2.
   a. What is the reason for identifying non-conforming products?

   b. What is the role of training in the implementation of ISO 9000?

   c. What is the purpose of calibrating equipment?

   d. What should be the organization's approach regarding customer orders and contracts.

   e. What should the organization's approach be when a supplier does not deliver goods or services as agreed?

   f. What is the function of the Quality Manual?

   g. Why identify the inspection and test status of a product?

   vii. To keep them under control, so that appropriate action such as rework can be taken.

   i. To ensure that everyone involved is capable of meeting the standard of work required.

   ii. To ensure standards of measurement are consistent.

   vi. To make sure they are properly understood, and that the organization is capable of meeting the customer's requirements.

   iv. To take action to prevent it happening again.

   v. To set out all the procedures for implementing the quality system.

   iii. To ensure that it has been verified and whether or not it conforms.

Reflect and review

Self assessment 3 on page 49

1. Total = 900, so mean = $\dfrac{\text{total}}{\text{number of value}} = \dfrac{900}{20} = 45$

    Range = 98 − 7 = 91

2.

Step	Calculation
	Data: the typing speed of five typists in words per minute: 80, 57, 72, 48, 63
1 Work out the mean of the set of values.	Total 320. Mean = $\dfrac{320}{5}$ = 64 wpm
2 Subtract the mean from each value, to give the 'differences'.	80 − 64 = 16 57 − 64 = −7 72 − 64 = 8 48 − 64 = −16 63 − 64 = −1
3 Take each of these differences and square it: multiply it by itself.	16 × 16 = 256 −7 × −7 = 49 8 × 8 = 64 −16 × −16 = 256 −1 × −1 = 1
4 Add up these squares.	256 + 49 + 64 + 256 + 1 = 626
5 Divide this sum by the number of items. The result is called the 'variance'.	$\dfrac{626}{5}$ = 125.2
6 Take the square root of the variance; this gives the standard deviation.	$\sigma = \sqrt{125.2}$ = 11.19 wpm (approx.)

3. Compare your sketch with the one shown below:

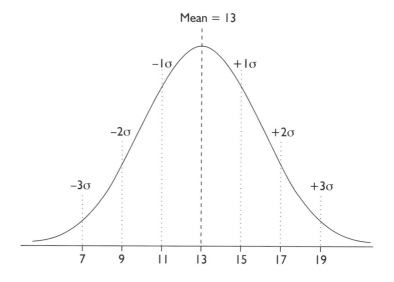

Reflect and review

4   The answers are:
    a   C and E
    b   C and F, or B and E
    c   C and G, or A and E.

Self assessment 4 on page 78

1   a   There are three discs with incorrect labels, so the probability of finding one of these is $3/20 = 0.15$.
    b   There are six faulty discs, so the probability of finding one of them is $6/20 = 0.3$.
    c   To work out the probability of one action or another, we add the probabilities: $0.3 + 0.3 = 0.6$.
    d   Strictly speaking, there is no way to be sure of finding all the faults. However, 100 per cent inspection will increase the chances considerably.

2   According to the table, for a box (or batch) size of 480, 32 samples should be taken. If any defects are found in the sample, the box should be rejected. Otherwise, it should be accepted.

3   a   The total of the values = 130. The mean of sample means = $130/10 = 13$. The standard deviation = 0.951.
    b   A suitable graph is shown below:

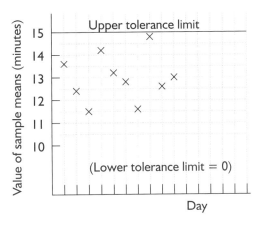

   c   We don't have a lower tolerance limit, so we can't simply apply the formula. The question is: 'Is the mean plus three standard deviations less than the upper tolerance limit?'

Mean + $3\sigma = 13 + (3 \times 0.951) = 15.853$. But the upper tolerance limit is fifteen minutes, so this work process is not capable. (Somehow, the service must be speeded up, or we can expect late deliveries – sooner or later!)

5 Answers to activities

Activity 3 on page 5

The answer is 'All of the above': everyone in the company must take responsibility for quality. Quality is not something that can be added on at the end of the line.

Reflect and review

Activity 18 on page 44

As you will see from the next diagram, the normal curve moves along the axis as the mean changes.

> This is what three sets of data with the same 'spread', but with different means, looks like.

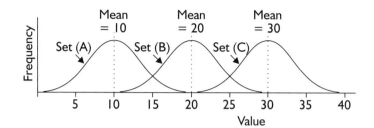

Activity 27 on page 59

The correct answer is that the letter A appears 49 times.

Activity 36 on page 70

a The total of the values = 70, so the mean of sample means = 7. The standard deviation = 3.435.

A suitable graph is shown below:

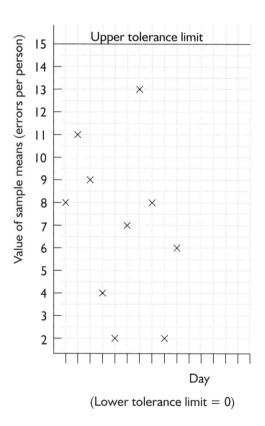

(Lower tolerance limit = 0)

b We don't have a lower tolerance limit, so we can't simply apply the formula. The question is: 'Is the mean plus three standard deviations less than the upper tolerance limit?'

Mean + 3σ = 7 + (3 × 3.435) = 17.305. But the upper tolerance limit is ten errors per person, so this work process is not capable of meeting the management target.

Reflect and review

6 Answers to the quick quiz

Answer 1   The starting point for quality is in the needs of the customer.

Answer 2   A **producer** generates goods or services for sale. A **supplier** offers goods or services for sale.

Answer 3   **Design quality** is the degree to which the specification of the product satisfies customers' wants and expectations. The **quality of conformance** is the degree to which the product conforms to specifications, when it is transferred to the customer.

Answer 4   Everybody!

Answer 5   A manager needs to behave as if he or she cares about quality, and has a commitment to it, so encouraging the rest of the staff to follow suit.

Answer 6   Continuous improvement means carrying out many detailed improvements to products, procedures and practices, over a long period.

Answer 7   The organization with ISO 9000 accreditation has proved that its quality system is under control, so that it is able to ensure delivery of products meeting customers' requirements.

Answer 8   A quality manual contains all the procedures for implementing the quality system.

Answer 9   An organization wishing to deliver quality products must have control over its suppliers. Should bought-in goods and services be faulty, the supplier must be traced so that corrective action can be taken, and to ensure that the problem does not recur.

Answer 10 Yes: 100 per cent inspection is not required.

Answer 11 The graph of a normal distribution shows a characteristic bell-shaped curve, symmetrical about its mean, and whose width depends on the standard deviation of the data. All large populations will tend to have graphs of this type.

Answer 12 a 1.0
           b zero.

Answer 13 Acceptable quality level (AQL) is the maximum percentage of defectives in a sample, that can be considered acceptable as a process average.

Answer 14 Process capability is the ability of a work process to produce output within a desired tolerance for a sustained period.

Answer 15 A **tolerance limit** is the outer boundary of tolerable values for a process. **Control limits** are set within tolerance limits to give an early indication that a process may be drifting out of tolerance.

Reflect and review

7 Certificate

Completion of this certificate by an authorized person shows that you have worked through all the parts of this workbook and satisfactorily completed the assessments. The certificate provides a record of what you have done that may be used for exemptions or as evidence of prior learning against other nationally certificated qualifications.

Pergamon Open Learning and NEBS Management are always keen to refine and improve their products. One of the key sources of information to help this process are people who have just used the product. If you have any information or views, good or bad, please pass these on.

NEBS MANAGEMENT DEVELOPMENT
SUPER SERIES
THIRD EDITION

Achieving Quality

....

has satisfactorily completed this workbook

Name of signatory ....

Position ....

Signature ....

Date ....

Official stamp

SUPER SERIES

SUPER SERIES 3

0-7506-3362-X Full Set of Workbooks, User Guide and Support Guide

A. Managing Activities

ISBN	Title
0-7506-3295-X	1. Planning and Controlling Work
0-7506-3296-8	2. Understanding Quality
0-7506-3297-6	3. Achieving Quality
0-7506-3298-4	4. Caring for the Customer
0-7506-3299-2	5. Marketing and Selling
0-7506-3300-X	6. Managing a Safe Environment
0-7506-3301-8	7. Managing Lawfully - Safety, Health and Environment
0-7506-37064	8. Preventing Accidents
0-7506-3302-6	9. Leading Change

B. Managing Resources

ISBN	Title
0-7506-3303-4	1. Controlling Physical Resources
0-7506-3304-2	2. Improving Efficiency
0-7506-3305-0	3. Understanding Finance
0-7506-3306-9	4. Working with Budgets
0-7506-3307-7	5. Controlling Costs
0-7506-3308-5	6. Making a Financial Case

C. Managing People

ISBN	Title
0-7506-3309-3	1. How Organisations Work
0-7506-3310-7	2. Managing with Authority
0-7506-3311-5	3. Leading Your Team
0-7506-3312-3	4. Delegating Effectively
0-7506-3313-1	5. Working in Teams
0-7506-3314-X	6. Motivating People
0-7506-3315-8	7. Securing the Right People
0-7506-3316-6	8. Appraising Performance
0-7506-3317-4	9. Planning Training and Development
0-75063318-2	10. Delivering Training
0-7506-3320-4	11. Managing Lawfully - People and Employment
0-7506-3321-2	12. Commitment to Equality
0-7506-3322-0	13. Becoming More Effective
0-7506-3323-9	14. Managing Tough Times
0-7506-3324-7	15. Managing Time

D. Managing Information

ISBN	Title
0-7506-3325-5	1. Collecting Information
0-7506-3326-3	2. Storing and Retrieving Information
0-7506-3327-1	3. Information in Management
0-7506-3328-X	4. Communication in Management
0-7506-3329-8	5. Listening and Speaking
0-7506-3330-1	6. Communicating in Groups
0-7506-3331-X	7. Writing Effectively
0-7506-3332-8	8. Project and Report Writing
0-7506-3333-6	9. Making and Taking Decisions
0-7506-3334-4	10. Solving Problems

SUPER SERIES 3 USER GUIDE + SUPPORT GUIDE

ISBN	Title
0-7506-37056	1. User Guide
0-7506-37048	2. Support Guide

SUPER SERIES 3 CASSETTE TITLES

ISBN	Title
0-7506-3707-2	1. Complete Cassette Pack
0-7506-3711-0	2. Reaching Decisions
0-7506-3712-9	3. Managing the Bottom Line
0-7506-3710-2	4. Customers Count
0-7506-3709-9	5. Being the Best
0-7506-3708-0	6. Working Together

To Order - phone us direct for prices and availability details
(please quote ISBNs when ordering)
College orders: 01865 314333 • Account holders: 01865 314301
Individual purchases: 01865 314627 (please have credit card details ready)

We Need Your Views

We really need your views in order to make the Super Series 3 (SS3) an even better learning tool for you. Please take time out to complete and return this questionnaire to Tessa Gingell, Pergamon Open Learning, Linacre House, Jordan Hill, Oxford, OX2 8BR.

Name: ....

Address: ....

....

Company & Position (if applicable): ....

Title of workbook: ....

If applicable, please state which qualification you are studying for. If not, please describe what study you are undertaking, and with which organisation or college:

....

Please grade the following out of 10 (10 being extremely good, 0 being extremely poor):

Content .... Appropriateness to your position ....

Readability .... Qualification coverage ....

What did you particularly like about this workbook?

Are there any features you disliked about this workbook? Please identify them.

Are there any errors we have missed? If so, please state page number: ....

How are you using the material? For example, as an open learning course, as a reference resource, as a training resource etc.

....

How did you hear about Super Series 3?:

Word of mouth: Through my tutor/trainer: Mailshot:

Other (please give details): ....

Many thanks for your help in returning this form.